Synthesis and Technique
in Inorganic Chemistry

A LABORATORY MANUAL

THIRD EDITION

Synthesis and Technique in Inorganic Chemistry

A LABORATORY MANUAL

Gregory S. Girolami and Thomas B. Rauchfuss

UNIVERSITY OF ILLINOIS AT URBANA–CHAMPAIGN

Robert J. Angelici

IOWA STATE UNIVERSITY

University Science Books
Sausalito, CA

University Science Books
55D Gate Five Road
Sausalito, CA 94965
Fax: (415) 332-5393
www.uscibooks.com

Production manager: *Susanna Tadlock*
Manuscript editor: *Jeannette Stiefel*
Designer: *Robert Ishi*
Illustrator: *John Choi*
Compositor: *Asco Typesetters, Hong Kong*
Printer and Binder: *The Maple-Vail Book Manufacturing Group*

This book is printed on acid-free paper.

Library of Congress Cataloging-in-Publication Data

Girolami, Gregory S., 1956–
   Synthesis and technique in inorganic chemistry : a laboratory
manual, — 3rd ed. / Gregory S. Girolami, Thomas B. Rauchfuss,
and Robert J. Angelici.
     p. cm.
  Rev. ed. of: Synthesis and technique in inorganic chemistry /
Robert J. Angelici, 2nd ed. c1986.
  Includes bibliographical references and index.
  ISBN 0-935702-48-2 (alk. paper)
  1. Chemistry, Inorganic—Laboratory manuals. I. Rauchfuss,
Thomas B., 1949– . II. Angelici, Robert J. III. Angelici, Robert
J. Synthesis and technique in inorganic chemistry. IV. Title.
QD155.G57 1998
542—dc21   98-14491
                                        CIP

Printed in the United States of America
10 9 8 7 6 5 4 3 2

Contents

Part I. SOLID-STATE CHEMISTRY

Part II. MAIN GROUP CHEMISTRY

Part III. COORDINATION CHEMISTRY

Part IV. ORGANOMETALLIC CHEMISTRY

Part V. BIOINORGANIC CHEMISTRY

Preface to the Third Edition

As in the two previous editions, the purpose of this laboratory manual is to instruct students in the modern synthetic and instrumental techniques currently used in inorganic chemistry. Our intent in this edition is to reflect the changes that have occurred in the field of inorganic chemistry in the last 20 years. We felt that it was important to place greater emphasis on the applications of inorganic synthesis in materials science and in biochemistry, and so we have introduced several experiments that illustrate these themes. We also felt, for a variety of reasons, that it was time to retire several of the experiments. Consequently, 15 of the 23 experiments in this edition are largely or entirely new. Some may miss their "old favorites", but we have striven to ensure that the new experiments work just as well (or better!), and that they provide opportunities to explore the expanding variety of inorganic chemical systems and characterization techniques. For those experiments that have been retained in this edition, the procedures have been improved and the references have been updated.

This edition retains many of the features that distinguished its earlier incarnations. In particular, we have continued the practice of closely associating synthesis and characterization methods. Almost every experiment highlights a particular technique in addition to the preparative details.

Nevertheless, we have made some changes in the structures of the experiments. In the previous editions, all of the experiments were designed so that they could be completed in one 3- or 4-hour laboratory period. While this approach has its conveniences, it also does not permit much exploration of the further chemistry of the compounds prepared. Students can lose sight of the important lesson that the synthesis of a compound is more often a means to an end rather than an end in itself. Therefore, we have included some experiments that can be extended to a second or third laboratory period, during which time students can investigate the reactivity of a compound they have just synthesized. In some cases, the first day's work merely involves setting up a reaction that will run overnight; the rest of the first laboratory period can be devoted to completing an experiment from the previous week. In other cases, the first day's work can be omitted: for example, in Experiment 23 the preparation of H_2TPP can be dispensed with if this material is on hand (it is commercially available but moderately expensive), and the synthesis of Cu(TPP) can be started right away. In addition, the instructor can often omit the second day's work if this is more convenient or desirable; the place where the first day's work ends is clearly marked. These options should provide the instructor with considerable flexibility in designing a course around this book.

At the end of each experiment, the section called "Independent Studies" provides ideas for more research-type investigations. The intent is to have the student extend the techniques learned in the experiment to other related systems. In selecting an independent study, the student must consult the literature to determine what chemicals and equipment are required, plan the experiment, and evaluate the results independently. In our experience, this has been a very worthwhile aspect of the course, challenging the student to undertake his or her own project. As in any research, many students will be successful in achieving their goal, while others will not; irrespective of the outcome, most students feel that the independent projects are the most memorable and exciting part of the course. The number of independent studies that can be carried out during the course will depend on the time available; in our one-semester course at Illinois we have generally devoted 12 hours of "in-lab" time to the independent projects.

We are indebted to John Arnold, John McDevitt, Robert Morris, Ken Poeppelmeier, and John Shapley for providing write-ups of new experiments that we have modified and incorporated into this edition. We also thank John Corbett, Peter Dorhout, Peter Fox, Richard Keiter, Scott McKinley, James Priepot, Jack Selegue, and David Wigley, who tested some of the new experiments and provided valuable comments. We hope that this revised edition will serve to introduce young chemists to the exciting and vigorous field of synthetic inorganic chemistry.

Gregory S. Girolami
Thomas B. Rauchfuss

Preface to the First Edition

Interest and research in inorganic chemistry have expanded in recent years. This has been partially a result of the discovery of new and exotic classes of compounds such as the noble gas fluorides, boron hydrides, and metal clusters. Probably of more importance is the recognition that inorganic chemistry is fundamental to vast areas of organic, physical, and biochemistry. These developments have largely stemmed from the synthesis and the detailed characterization of new compounds.

Realizing that our students, upon graduation, would be doing laboratory work with the most modern of facilities in either graduate school or industry, we designed a junior-senior level inorganic laboratory course to prepare them for their future encounter with research. To be of value the course had to include techniques which are used in chemical research today. Most of the techniques which we think are important have been included in our course and in this manual:

a. Inorganic synthetic techniques—for example, those requiring a dry box or bag, a vacuum line, very high temperatures, nonaqueous solvents, and electrolytic oxidation.
b. Methods of purification—ion exchange, thin layer and column chromatography, vacuum sublimation, distillation, recrystallization, and extraction.
c. Methods of characterization—infrared, ultraviolet-visible and nuclear magnetic resonance spectroscopy, mass spectrometry, ionic conductivity, optical rotations, magnetic susceptibilities, chemical reactivity and rates of reaction, and equilibrium constants for complex formation.

The experiments have been designed to illustrate the fundamentals of these techniques. It is assumed that if refinements or slight modifications of these techniques are required in any future research, they can be made by consulting the specialized references on techniques given at the end of each experiment.

The selection of experiments has been governed by several practical considerations. First, inherently hazardous procedures have been avoided, although lack of understanding or ignorance of proper methods can make any of the experiments dangerous. Thus careful attention should be paid to the safety notes given in the instructions. Second, for those institutions which have scheduled 3-

hour laboratory periods the experiments have been selected to allow experimental work in 3-hour segments. The experiments which require longer periods are noted in the text. Third, we have limited the preparations to those which require relatively inexpensive chemicals. Finally, we felt that it was highly desirable for the student to prepare compounds which are either not commercially available or are very expensive. This again is an attempt to place the student in the position of a research chemist who would obviously buy needed compounds if they could be purchased at reasonable prices from commercial sources. This means that the syntheses in this manual are largely of compounds which were only recently prepared and which border on the frontiers of research.

In pursuing the above goals, the experiments have necessarily involved preparations and characterizations of a variety of types of compounds. Hence syntheses of nonmetal, transition metal, and organometallic compounds are represented. The experiments are all written in considerable detail so that the student may learn new techniques properly and as rapidly as possible. Many of the results, such as infrared and nuclear magnetic resonance spectra, optical rotations, mass spectra, and magnetic susceptibilities, will require interpretation. This interpretation will lead the student to the library to examine original papers and standard reference works. This would again place the student in the shoes of a researcher, and indeed it would be very desirable, if time permits, to extend his research to carrying out a synthesis which he himself would select from the literature.

During the past four years at Iowa State, this one-quarter course has involved two three-hour laboratory periods each week. Since the course is offered at the junior-senior level, all students had completed physical, analytical, and organic chemistry laboratory courses and were taking an independent lecture course in inorganic chemistry concurrently. Essentially the same course was available for first-year graduate students. During the one-quarter term, it was possible to complete approximately eight of the experiments. Since facilities vary from one institution to another and very few institutions will have equipment to do all the experiments given in this manual, a sufficient range of experiments is provided to create a meaningful course at almost all colleges and universities. Because the lack of nuclear magnetic resonance and mass spectral facilities is a very common problem, these spectra have been included at the back of the book for compounds which are amenable to characterization by these techniques. These can then be interpreted by the student. If NMR and mass spectral equipment are available, it would of course be more valuable to have the student obtain a spectrum on his own preparation.

In general, there is no required order for performing the experiments. To make the most efficient use of our facilities it was necessary to have each student perform a different experiment during a given laboratory period. Hence, experiments from throughout the manual were being carried out even during the first week of the course. The only restriction on the ordering of experiments results from the dependence of some measurements upon a previous preparation, such as the kinetic study of the aquation of $Co(NH_3)_5Cl^{2+}$ to $Co(NH_3)(OH_2)^{3+}$ which requires the prior synthesis of $Co(NH_3)_5Cl^{2+}$. In this case and others, some of the necessary compounds may be purchased commercially if desired. Hints on the organization of the laboratory and on commercial sources for special equipment and chemicals are given in Notes to the Instructor.

Our experience with this laboratory course has been a particularly exciting and gratifying one. It is in large measure the enthusiasm of the students which has led to the writing of this book. For suggestions and considerable help in the development of the experiments, I am very much indebted to J. Graham, R. Bertrand, G. McEwen, Dr. J. Espenson, and especially D. Allison and D. White. For the typing of the experiments, I am grateful to Mrs. L. Gustafson, Mrs. P. Feikema, and Mrs. L. Dayton. My gratitude goes to Forrest Hentz at North Carolina State University and Robert Kiser at the University of Kentucky for their reviews and suggestions on the final manuscript. Finally, I want to thank Iowa State University for providing me with the opportunity to develop and organize the experiments in this book.

Robert J. Angelici

Synthesis and Technique in Inorganic Chemistry

A LABORATORY MANUAL

Introduction

This book is designed to be an introduction to modern research techniques in inorganic chemistry. The experiments to be carried out involve the synthesis of various types of compounds by diverse experimental techniques. Modern instrumental methods will be used to characterize the products. Inorganic compounds are crucial to many industrial and biochemical processes, and we hope this book will convince you that the synthesis of inorganic compounds is both important and exciting.

This first chapter will expand upon certain aspects of (1) safety, (2) basic laboratory procedures, and (3) methods of keeping a research notebook. You must take care to acquaint yourself with the safety hazards related not only to the chemical properties of reagents but also to the dangers presented by unfamiliar apparatus. New fundamental techniques will be encountered, and these must become as second nature as weighing was in prior courses. Finally, the research orientation of the course requires careful record-keeping by the researcher. It is important that you become well acquainted with this material before beginning the experiments.

SAFETY

The importance of adequate safety precautions cannot be overemphasized. Because the chemicals and equipment used in this course can be dangerous if handled improperly, it is very important that you be completely familiar with the experiment before beginning it. Read the experimental directions critically before coming to the laboratory, and investigate the toxicity and reactivity of the chemicals used and the hazards related to the apparatus. This knowledge permits you to determine what is and what is not a dangerous situation. Throughout this book safety precautions are mentioned for your protection; pay particular attention to these warnings. *When undertaking any hazardous operation, know beforehand what you would do in case of an accident.*

Eye Protection. *Safety goggles should be worn at all times in the laboratory.* Safety goggles protect you not only from your own experiment but also from your neighbor's, and therefore they are preferred over safety glasses. For a particularly hazardous operation, a face shield should be used, whereas flasks under vacuum and sealed-tube reactions should be placed behind a portable safety shield. If an accident does occur, and a chemical is splashed in the eyes or brought to the eyes by contaminated fingers, immediately wash the eyes with a large amount of water, using a hose attached to a water faucet or an eyewash fountain. Make sure that your instructor knows of the situation immediately.

Clothing. A laboratory coat should always be worn, and rubber or polyethylene gloves should be used when handling corrosive liquids. Open-toed shoes, short pants, and excessively loose or flowing clothes are forbidden. Long hair should be tied back.

Fire and First Aid Equipment. In case of fire, you should be familiar with the location and operation of fire extinguishers. Usually, a pin must be removed from the handle of the extinguisher before it can be used. Take care not to hold the nozzle of the extinguisher too close to the fire: the burst of foam or powder can spread a fire rather than extinguish it. Liquid nitrogen is also a good fire extinguisher. Immediately inform your instructor if a fire occurs.

A fire blanket should be on hand to be wrapped around persons whose clothing catches fire. The location of safety showers and eyewashers should be known and their use demonstrated. To use the safety shower, stand beneath it and pull down on the lever. A well-stocked first aid kit should also be available (see references for recommended kit contents).

Chemical Toxicities. *Always assume that chemicals are toxic unless you know otherwise.* Solids are relatively harmless because they can only be taken into the body orally or through open cuts on the hands and arms. On the other hand, vapors of volatile liquids and gases are much more dangerous, and their presence is frequently difficult to detect. For this reason, *smoking, eating, or drinking in the laboratory are prohibited.* In addition, use gloves when necessary, and wash your hands after each laboratory period. Any reaction that involves or produces toxic fumes should be performed in a hood.

Federal law requires that sellers of chemicals provide to the purchaser information about the known health, fire, explosion, and reaction hazards of a chemical. This information is collected in a materials safety data sheet, or MSDS, which also contains sections on how to handle the compound, what first aid procedures should be followed in case of contact or inhalation, and how to dispose of the chemical safely. These sheets should be kept on file by your chemistry department, and should be available upon request. If you are unfamiliar with the hazards of any chemical, consult the appropriate MSDS sheet or the references at the end of this section.

The toxicities of certain common chemicals such as nitrobenzene, bromine, and carbon tetrachloride are surprisingly high. Table I-1 lists threshold limit values (TLV) in parts per million (ppm) of a variety of substances in air at 25 °C. Threshold limit values are concentrations of gas to which nearly all workers could be exposed for periods of months without adverse effect. Somewhat higher concentrations could be experienced for shorter periods. Unfortunately, most research laboratories do not have facilities for measuring concentrations of toxic gases in the air. For this reason, the numbers in Table I-1 serve primarily to indicate relative toxicities. For example, the exceedingly toxic hydrogen cyanide, HCN, has a TLV of 4.7, which suggests that other substances with values of 5 or less, such as AsH_3, B_2H_6, F_2, $Fe(CO)_5$, $Ni(CO)_4$, NO_2, and SO_2, will also be very toxic. Moreover, these substances are highly volatile and are likely to give high gas concentrations. On the other hand, chemicals such as ethyl acetate

Table I-1
Threshold Limit Values of Common Gases and Volatile Liquids[a]

	TLV (in ppm)		TLV (in ppm)
Acetic acid	10	Ethylamine	5
Acetic anhydride	5	Ethylenediamine	10
Acetone	750	Fluorine	1
Acetonitrile	40	Formaldehyde	0.3
Ammonia	25	Germane	0.2
Aniline	2	n-Hexane	50
Arsine	0.05	Hydrazine	0.1
Benzene	10	Hydrogen bromide anhydrous	3
Benzaldehyde	2	Hydrogen chloride anhydrous	5
Benzoyl chloride	0.5	Hydrogen cyanide	4.7
Benzyl chloride	1	Hydrogen fluoride anhydrous	3
Bromine	0.1	Hydrogen peroxide 90%	1
Bromoethane (ethyl bromide)	5	Hydrogen selenide	0.05
Bromomethane (methyl bromide)	5	Hydrogen sulfide	10
1,3-Butadiene	2	Iodine	0.1
2-Butanol	100	Iodomethane (methyl iodide)	5
n-Butylamine	5	Iron pentacarbonyl	0.1
Carbon disulfide	10	Mercury	0.02
Carbon monoxide	25	Mesitylene (1,3,5-trimethylbenzene)	25
Carbon tetrachloride	5	Methanethiol (methyl mercaptan)	0.5
Chlorine	0.5	Methanol (methyl alcohol)	200
Chlorobenzene	10	Nickel carbonyl	0.05
Chloroethane	1000	Nitric acid	2
Chloroform	10	Nitric oxide	25
Cyclohexane	300	Nitrobenzene	1
Cyclohexene	300	Nitrogen dioxide	3
Cyclopentadiene	75	Nitromethane	20
Decaborane	0.05	Phenol	5
Diborane	0.1	Phosgene (carbonyl chloride)	0.1
o-Dichlorobenzene	25	Phosphine	0.3
p-Dichlorobenzene	10	Phosphorus trichloride	0.2
1,2-Dichloroethane (ethylene chloride)	10	Propanoic acid (propionic acid)	10
Dichloromethane (methylene chloride)	50	Pyridine	5
Dicyclopentadiene	5	Sulfur monochloride	1
Diethylamine	5	Sulfur dioxide	2
Diethyl ether	400	Sulfur hexafluoride	1000
N,N-Dimethylformamide	10	Sulfur pentafluoride	0.01
Dimethyl sulfate	0.1	Tetrahydrofuran	200
p-Dioxane	25	Toluene	50
Ethyl acetate	400	Triethylamine	1
Ethanol (ethyl alcohol)	1000	p-Xylene	100

[a] Data available May, 1998 from the US Department of Labor Occupational Safety and Health Agency: http://www.osha-slc.gov/OCIS/toc_chemsamp.html

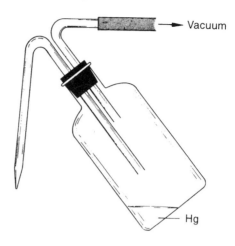

Figure I-1
Vacuum flask for cleaning up mercury spills.

are relatively harmless because of their large TLV (400) and also their lower volatility.

Because chemical research involves the preparation of new compounds whose toxic effects are totally unknown, it is best to establish safe laboratory practices based on the assumption that all chemicals are toxic, particularly those that vaporize readily. If volatile liquids and gases are always handled carefully in an efficient hood, their toxicity should never produce a dangerous situation. *When in doubt about the toxicity of reactants or products in a reaction, always perform the reaction in a hood.*

Mercury is commonly used in the inorganic laboratory and is employed in several of the experiments in this book. At 25 °C, it has a vapor pressure of 1.7×10^{-3} Torr, which would be very toxic if it reached that pressure in a room. Fortunately, room ventilation usually keeps the actual level in a room far below that pressure. Mercury should always be stored in tightly closed bottles and never allowed to spill. Spilled mercury easily hides in corners and under stationary objects, providing an undesirable source of mercury vapor. Immediately inform your instructor when a mercury spill occurs or is noticed. Once mercury has spilled, it is best removed by sucking it up with the vacuum flask shown in Figure I-1. This apparatus consists of a 6-mm glass tube drawn to a 1-mm tip opening, which is fitted to a trap connected to a water aspirator or vacuum pump. As much of the spilled mercury as possible should be collected with this apparatus, although it is a tedious chore. Afterward, the place where the spill occurred can be sprinkled with powdered sulfur. The remaining invisible mercury droplets react slowly with sulfur to form mercuric sulfide, (HgS). Because the formation of HgS occurs only on the surface of the droplets and a disturbance of the droplet produces a fresh mercury surface, treatment with sulfur is at best a temporary method of reducing the vaporization of mercury. By far the best method of avoiding mercury contamination in a laboratory is not to spill it.

Other Chemical Hazards. Chemicals present dangers besides those attributed to their toxicities. All organic solvents should be regarded as inflammable; open flames are forbidden in the laboratory except under the direct supervision of an instructor.

Some chemicals produce severe burns when they come in contact with the skin. These should be handled with gloves, and an antidote should be available (e.g., sodium thiosulfate for a Br_2 burn). If a corrosive chemical does get on your skin, wash the area thoroughly with water and then treat it with an antidote. For acids and bases, follow the water wash with a solution of sodium bicarbonate in water. If a large spill occurs that cannot be quickly washed off under the faucet, use a safety shower. Report any spill to your instructor immediately.

Pipetting by mouth is strictly prohibited. Use a suction bulb instead, preferably one that has "squeeze" valves. Do not smell a chemical directly, and do not handle chemicals with your bare hands.

Acid and Base Baths. Various corrosive solutions are sometimes used to assist in cleaning glassware. Chromic acid should not be used because it is toxic and carcinogenic. Aqua regia (a mixture of nitric acid and hydrochloric acid) is also not recommended because it evolves the toxic gases chlorine and nitrosyl chloride ($NOCl$). If you must use an acid bath, we recommend the Nochromix formulation available from Aldrich. Ethanolic potassium hydroxide is caustic but useful for degreasing.

Both acid and base baths can cause extreme burns. If you use any of these solutions, be sure to wear arm-length rubber gloves. Acid baths are potent oxidizers, and you should *never* expose them to organic solvents such as acetone or organic residues because an explosion can take place. Before being introduced into any cleaning bath, glassware should first be scrubbed with a brush, soap, and water to remove as much residue as possible, and then rinsed with water.

Heating Mantles and Baths. For most laboratory experiments, heating mantles are preferred over heating baths because they are safer to use. The principal disadvantage of heating mantles is that the heating is less even and "hot spots" can sometimes lead to pyrolysis of the reaction mixture. If you must use mineral oil and paraffin baths, they should *never* be heated above about 150 °C. They can easily catch fire; mineral oil baths will smoke near their flash point, so immediately discontinue heating if smoke appears. Another danger of oil baths can arise if water gets into them: Above 100 °C, water droplets on the bottom of the bath will flash evaporate, throwing hot oil for a considerable distance. *Always* check oil baths for the presence of water; if you cannot see through to the bottom of the bath, the oil should be discarded. If you require temperatures above 150 °C, you can use a silicone oil bath (max temp 250 °C), a Woods' metal bath (max temp 350 °C), a sand bath (max temp 350 °C), or preferably a heating mantle (max temp 350 °C). Open flames are strictly forbidden.

Potential Explosion Hazards. *Never* heat a closed system. Pressure can build up and cause an explosion. Place an appropriate shield between you and a potentially explosive reaction mixture. Such explosive situations have been avoided in this course as much as possible, but when glass flasks are evacuated, they may implode with serious consequences. Vacuum-jacketed Dewars will also implode if they are broken. If Dewars are wrapped on the outside with sturdy tape, implosions are considerably less dangerous.

Ultraviolet Light Sources. Ultraviolet (UV) light is very harmful to the eyes. Never look directly at a UV lamp. For reactions that require irradiation with UV light for long periods, carry out the reaction in a hood with the window all the way down, and the window covered so that no light escapes.

Disposal of Chemicals. Metallic lithium, sodium, and potassium should be used with extreme caution owing to their violent reaction with water; similar precautions should be taken with alkyl lithium, Grignard, and alkyl aluminum reagents. All of these materials are severe fire hazards; they should never be placed in a waste bucket. Any reaction mixtures containing these materials should be cautiously treated first with isopropanol, then ethanol, and finally water. Trying to hydrolyze these reagents too quickly can lead to ignition or explosion of evolved gases.

Labeled waste containers should be available for chlorinated solvents, nonchlorinated solvents, and solid residues. If you have any doubts about the disposal of waste chemicals, consult your instructor. For a general reference, see *Prudent Practices in the Laboratory*, by the National Research Council.

REFERENCES

http://ull.chemistry.uakron.edu/erd/. A searchable index of hazardous chemicals.

http://www.osha-slc.gov/OCIS/toc_chemsamp.html. Toxicity data for many chemicals.

Breazeale, W. H. J. Jr.; Ramsey, H. *Chem. Eng. News* **1998**, *76* (22), 6. Study of the wearing of contact lenses in the chemical laboratory.

Budavari, S. M., Ed. *The Merck Index: An Encyclopedia of Chemicals, Drugs, and Biologicals*, 11th ed., Merck & Co: Rahway, NJ, 1989. The basic reference to the toxicities of a large number of chemicals.

Lenga, R. E., Ed. *The Sigma–Aldrich Library of Chemical Safety Data*, 2nd ed., Sigma–Aldrich: Milwaukee, WI, 1987. Information about chemicals found in the Aldrich–Fluka–Sigma catalog can be found on-line at http://www.sigma.sial.com.

McKusick, B. C. *J. Chem. Educ.* **1984**, *61*, A152. Procedures for laboratory destruction of chemicals.

Prudent Practices in the Laboratory, National Academy Press: Washington DC, 1995. An excellent reference covering the handling and disposal of laboratory chemicals.

Safety in Academic Chemistry Laboratories, 6th ed., American Chemical Society: Washington DC, 1995. A paperback guide that covers a broad spectrum of safety aspects.

Sax, N. I., Ed. *Dangerous Properties of Industrial Materials*, 7th ed., 3 Vols. Van Nostrand Reinhold: New York, 1988. Good general reference.

Steere, N. V., Ed. *Safety in the Chemical Laboratory*, 4 Vols. Chemical Education Publishing Co.: Easton, PA, 1964–1980. An excellent source of information and references on chemical safety, including a section on accident case histories.

Steere, N. V., Ed. *CRC Handbook of Laboratory Safety*, 2nd ed., CRC Press: Cleveland, OH, 1976. Good general reference.

LABORATORY PROCEDURES

Use of Time. The efficient use of time is an asset not only to a student but especially to a researcher. Plan your experiments so that you will profitably use time that you would otherwise spend watching something that needs no attention. This course allows some latitude in the planning of experiments, and you should be constantly looking for opportunities to use the available time effectively.

Cleanliness. Because most of the experiments will involve the use of equipment that other students will use sometime during the course, it is essential that all equipment be left in good condition at the end of each period. Any equipment that is broken should be reported to the instructor so that a replacement may be found in time for the next class. Glassware should be cleaned after use.

Compressed Gas Cylinders and Lecture Bottles. Several experiments in this book make use of gases that are commercially available in compressed gas cylinders or lecture bottles. Compressed gas cylinders should always be securely anchored to a wall or heavy bench with a cylinder clamp or chain. Free standing gas cylinders are extremely dangerous: If the cylinder should fall over and snap off the valve, the cylinder becomes a jet-propelled rocket that can easily go through brick walls. Lecture bottles are small pressurized gas cylinders about 35 cm long and 5 cm in diameter. They should be placed in a hood and secured with the help of a strong clamp fastened to a ring stand or other support capable of holding the lecture bottle firmly in place.

Gas cylinders come in a variety of sizes with several types of valves and regulators; the manufacturer's catalog should be consulted for the correct valve fittings. Also, the metallic content of the valves may be dictated by the corrosive properties of the gas. Many cylinders contain a safety valve or nut, which is designed to rupture if the pressure inside the cylinder exceeds its limit. Under no circumstances should you tamper with the safety nut.

The main valve (Fig. I-2) on a cylinder is simply an on–off valve that allows no control of the gas flow; it should always be used in conjunction with some type of control valve. A needle valve (Fig. I-2) permits such control, but if the cylinder

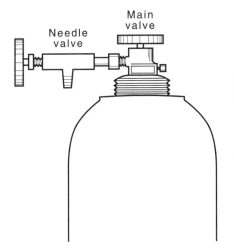

Figure I-2
Gas cylinder with attached needle value.

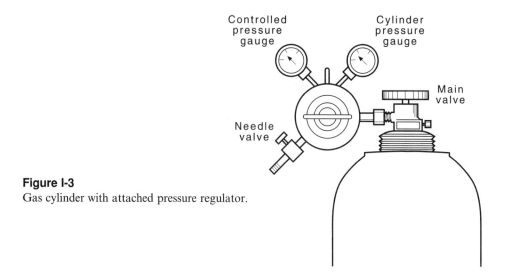

Figure I-3
Gas cylinder with attached pressure regulator.

contains a compressed gas, the cylinder pressure will decrease as the cylinder is used and the gas flow will likewise decrease. Thus, for compounds that exist as gases in the cylinder (e.g., H_2, O_2, CO, N_2, and Ar), a given flow rate cannot be maintained without continuous adjustment. Compounds that condense to form liquids under pressure exert their natural vapor pressure so long as any liquid remains in the cylinder. For these gases (e.g., NH_3, CO_2, and HCl), a continuous flow rate can be obtained with a needle valve.

To achieve a constant flow rate for gases that do not condense under the pressure in the cylinder, a pressure regulator (Fig. I-3) is required. First, open the main valve; the gas pressure in the cylinder is given on the right-hand gauge. Then open the regulator valve by turning the lever clockwise. Finally, adjust the flow rate to the desired level by opening the needle valve. The pressure between the needle valve and the regulator valve is given on the left-hand gauge. The regulator will maintain this pressure. During the experiment, the flow can be monitored by passing the gas through an oil bubbler after it exits from the reaction flask (Fig. I-4). When handling air sensitive compounds, as is done in several

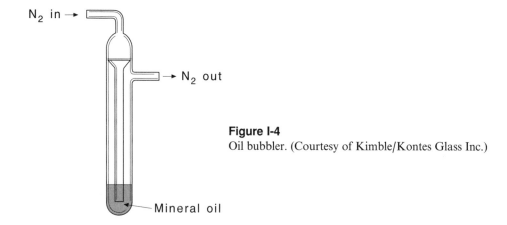

Figure I-4
Oil bubbler. (Courtesy of Kimble/Kontes Glass Inc.)

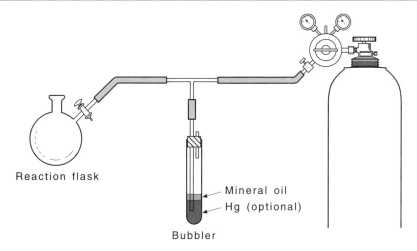

Figure I-5
Nitrogen gas supply equipped with mercury bubbler.

experiments in this book, it is good practice to monitor the "bubble rate" regularly to ensure that your sample is properly blanketed with a protective layer of inert gas. The flow can be halted by closing the needle valve, but when you are finished with the cylinder for the day, close the main valve to prevent loss of the gas in case the regulator leaks slightly. Do not empty a cylinder completely; leave approximately 200 kPa ($\sim$2 atm or 25 lb in^{-2}) in the cylinder so that it does not become contaminated with air or other gases before it is returned to the supplier for refilling.

In several experiments, N_2 gas will be used to flush air from a reaction system, as in Figure I-5. Before the reaction is begun, the stopcock on the reaction flask is often closed. To prevent any pressure buildup, which could result in the "popping" of the rubber tubing connecting the apparatus to the N_2 cylinder, it is convenient to connect a mercury bubbler to the rubber tubing. The bubbler acts as a vent for the excess N_2 (Fig. I-5). The bubbler is filled with mercury to a height of about 5 cm, and the bubbler should be vented to a fume hood. The mercury should also be covered with a layer of mineral oil to minimize vaporization of the toxic mercury.

Utility Vacuum Line. A simple vacuum line is very convenient for carrying out routine sublimations, vacuum distillations, and the vacuum drying of products. One safety hazard associated with the vacuum line is the trap, which is usually cooled with liquid N_2. The trap should *never* be closed off for more than a few seconds: It should either be open to the vacuum pump or open to the air. If the trap is left open to the air, the liquid N_2 Dewar should be removed to prevent O_2 from condensing, because liquid O_2 can explode with trapped-out compounds upon warming. If pale blue liquid O_2 is noticed in the trap, it should be removed as follows: Immediately raise the Dewar to cool the trap with liquid N_2, close any stopcocks open to air, and pump out the trap for several minutes.

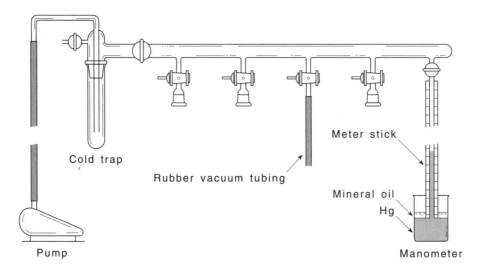

Figure I-6
Utility vacuum line.

A system that we have used is shown in Figure I-6. Comments on the construction and use of a vacuum line are given in Experiment 6 together with references to detailed accounts of vacuum line techniques. The system illustrated in Figure I-6 is first evacuated by closing all the stopcocks that open to the air and opening those that do not. Then a Dewar flask containing liquid N_2 is placed around the cold trap. Flasks to be evacuated may be connected to the vacuum through the rubber vacuum tubing. To evacuate the flask, open both the stopcock on the line and the one on the flask. If the vacuum line has "three-way" stopcocks, and an inert gas manifold, the flask can be filled with inert gas by turning the stopcock key on the vacuum line to the appropriate position.

Recrystallized materials that are still damp from the solvent may be dried by evaporating the solvent in a vacuum. The vessel containing the sample is attached to the line through either a ground glass joint or the rubber tubing. It is then opened to the vacuum and allowed to dry under a dynamic vacuum with the pump operating. The evaporated solvent collects in the cold trap and the solvent should be discarded when the drying is finished. See Experiment 6 for instructions on shutting down the vacuum line after use.

Make sure that any vessel you wish to evacuate is one that can withstand a vacuum, such as a round-bottom flask or tube; do not use Erlenmeyer flasks because they might implode. As a general rule of thumb, the largest volume that can be safely evacuated is a 1-L flask.

Because a fair degree of competence in glassblowing is required to construct a vacuum line, this topic will not be covered in this book. Following are several books that will be helpful in improving your glassblowing technique.

REFERENCES

Barbour, R. *Glassblowing for Laboratory Technicians*, Pergamon: London, 1968.

Braker, W.; Mossman, A. L. *Gas Data Book*, 6th ed., Matheson Gas Products: East Rutherford, NJ, 1980.

Hammesfahr, J. E.; Stong, C. L. *Creative Glassblowing*, Freeman: San Francisco, 1968.

Handbook of Compressed Gases, 3rd ed., Compressed Gas Association: Arlington, VA, 1990.

Shriver, D. F.; Drezdzon, M. A. *The Manipulation of Air-sensitive Compounds*, 2nd ed., Wiley: New York, 1986.

RESEARCH NOTEBOOK

The communication of experimental observations and their interpretation is an important duty of the scientist. Without it, little would be gained from the scientist's efforts. The first step in the communication chain is the accurate and detailed recording of experimental facts in a bound notebook. The purpose of this record is to allow you or someone else to learn from what you did in the experiment and to help you or others to repeat your success or avoid your failure. As you will soon learn in this course, detailed information about a synthesis or measurement is much appreciated by someone wishing to repeat your experiment. Your notebook record of your experiments should include more than enough detail to allow you or someone else to repeat the experiment successfully. It is much better to be overly detailed than to overlook observations that may be of use later.

If the experiment was unsuccessful, as some are, your notebook should contain sufficient detail about the experiment to allow you to make intelligent corrections in the procedure to increase your chance of success in the next attempt. For these reasons, your notebook should contain drawings of experimental apparatus (or a reference to a figure in this book; if the apparatus is not identical to that in the figure, alterations should be explicitly stated). Your notebook should also contain experimental observations such as volumes of liquids and weights of solids used, color changes, temperatures of reaction mixtures, reaction times, difficulties encountered, measurements, and cross references to spectra. Label spectra with the number of the notebook page on which the compound preparation is given.

All of these experimental details should be recorded in the notebook at the time of the observation. Do not record data on loose pieces of paper. This rule was not instituted by a cranky teacher; it is simply a waste of time and possible source of error to record observations and then recopy them into a notebook. If this rule is followed, your research notebook will not be particularly neat, but it should be readable. Because water and acid stains will contribute to the appearance of the notebook, it is important that your records be kept with permanent ink. Mistakes should be crossed out with a single line and a brief reason stated as to why the item was lined out.

Next, write down your conclusions concerning the experiment. Information specifically requested at the end of each experiment should also be included at this point. Each page of the notebook should be dated to indicate the day that the experiment was performed. To facilitate referring to experiments, the first two or three pages in the notebook should be left blank for a table of contents. As you complete an experiment, its title and page should be recorded in this table of contents.

The next step in communicating results to other scientists is to prepare a formal and orderly report stressing successful techniques used to achieve the desired goal and the general conclusions resulting from the study. Such a neat, well-organized report will probably not go beyond your instructor in this course, but in academic or industrial research such a record will be transmitted to other scientists in the company or university and very frequently throughout the world. Clear reports allow others to build upon your contributions.

REFERENCES

Dodd, J. S., Ed. *The ACS Style Guide: A Manual for Authors and Editors*, American Chemical Society: Washington DC, 1986.

Ebel, H. F.; Bliefert, C.; Russey, W. E. *The Art of Scientific Writing*, VCH: New York, 1990.

Kanare, H. M. *Writing the Laboratory Notebook*, American Chemical Society: Washington DC, 1985.

Schoenfeld, R. *The Chemist's English*, 2nd ed., VCH: New York, 1986. A series of enjoyable and useful essays on the art of writing clear scientific prose.

Strunk, W. Jr.; White, E. B. *The Elements of Style*, 3rd ed., Macmillan: New York, 1979. The definitive guide to good writing that should be on every writer's desk.

STANDARD REFERENCES TO SYNTHESES AND TECHNIQUES IN INORGANIC CHEMISTRY

Adams, D. M.; Raynor, J. B. *Advanced Practical Inorganic Chemistry*, Wiley: New York, 1965.

Armarego, W. L. F.; Perrin, D. D. *Purification of Laboratory Chemicals*, 4th ed., Butterworth–Heinemann: Oxford, UK, 1996. An excellent source of techniques useful for purifying a wide variety of solids and liquids.

Brauer, G. *Handbook of Preparative Inorganic Chemistry*, Vols. I and II, 2nd ed., Academic: New York, 1963. Contains many hundreds of useful recipes.

Eisch, J. J.; King, R. B. *Organometallic Syntheses*, Vol. 1–3, Academic: New York. Volume 1 contains excellent recipies for cyclopentadienyl and carbonyl complexes of the transition elements; Volume 2 contains a wide range of recipes for organometallic derivatives of the main group elements.

Errington, R. J., Ed. *Advanced Practical Inorganic and Metalorganic Chemistry*, Blackie: New York, 1997.

Inorganic Syntheses, Vols. 1–32, 1939–1998. Volumes 1–17 were published by McGraw-Hill and reprinted by Krieger; Volumes 18–32 were published by Wiley. An excellent source of reliable syntheses.

Jolly, W. L. *The Synthesis and Characterization of Inorganic Compounds*, Prentice-Hall: Englewood Cliffs, NJ, 1970.

Pass, G.; Sutcliffe, H. *Practical Inorganic Chemistry*, 2nd ed., Halsted Press: New York, 1974.

Schlessinger, G. G. *Inorganic Laboratory Preparations*, Chemical Publishing Co.: New York, 1974.

Szafran, Z.; Pike, R. M.; Singh, M. M. *Microscale Inorganic Chemistry*, Wiley: New York, 1991.

Woolins, J. D., Ed. *Inorganic Experiments*, VCH: New York, 1994.

STANDARD TEXTBOOKS AND REVIEWS OF INORGANIC CHEMISTRY

Abel, E. W.; Stone, F. G. A.; Wilkinson, G., Eds. *Comprehensive Organometallic Chemistry II*, Pergamon: New York, 1995. A thorough review of recent work in organometallic chemistry in 14 volumes.

Bailar, J. C. Jr.; Emeleus, H. J.; Nyholm, R.; Trotman-Dickenson, A. P., Eds. *Comprehensive Inorganic Chemistry*, Pergamon: Oxford, UK, 1973. A series of five volumes covering virtually all areas of inorganic chemistry.

Cheetham, A. K.; Day, P., Eds. *Solid State Chemistry: Techniques,* Oxford University: Oxford, UK, 1987.

Cheetham, A. K.; Day, P., Eds. *Solid State Chemistry: Compounds,* Oxford University: Oxford, UK, 1992.

Cotton, F. A.; Wilkinson, G. *Advanced Inorganic Chemistry*, 5th ed., Wiley: New York, 1988. Although this text is not always mentioned at the end of each experiment, it is an excellent source of further information and references on the chemistry illustrated in many of the experiments.

Gmelin's Handbook of Inorganic Chemistry, Verlag Chemie: Berlin. A comprehensive multivolume series covering all of inorganic chemistry; earlier volumes were published in German. An excellent source for the older literature, but more recent work (i.e., in the last 20 years) is often not included in the supplementary volumes.

Greenwood, N. N.; Earnshaw, A. *Chemistry of the Elements*, 2nd ed., Pergamon: New York, 1997. A textbook of inorganic chemistry; contains an excellent overview of main group chemistry.

King, R. B., Ed. *Encyclopedia of Inorganic Chemistry*, Wiley: New York, 1994. An eight-volume reference work containing many useful articles.

Wells, A. F. *Structural Inorganic Chemistry*, 5th ed., Oxford University: Oxford, UK, 1984. An authoritative text on solid-state inorganic chemistry.

Wilkinson, G.; Gillard, R. D.; McCleverty, J. A., Eds. *Comprehensive Coordination Chemistry*, Pergamon: New York, 1987. A thorough review of the field of coordination chemistry in seven volumes.

USEFUL LINKS ON THE WORLD WIDE WEB

http://www.library.ucsb.edu/subj/chemistr.html. A well-maintained collection of links to chemistry web sites.

http://www.shef.ac.uk/chemistry/chemdex/. A collection of thousands of links to chemically-oriented web sites.

Part I

SOLID-STATE CHEMISTRY

The 1-2-3 Superconductor YBa$_2$Cu$_3$O$_7$

Note: This experiment requires 6 hours spread over two laboratory periods.

In this experiment, you will prepare the solid-state compound YBa$_2$Cu$_3$O$_7$ by the reaction of the oxides of yttrium, barium, and copper in a tube furnace. Few compounds in recent history have generated as much excitement as this one. The intense interest in YBa$_2$Cu$_3$O$_7$ arises from the observation that it becomes a superconductor—a material that conducts electrical current with zero resistance. Superconductors are used in the construction of powerful magnets and may in the future be used to carry electrical power over long distances with 100% efficiency. Most solids that become superconducting are of limited utility because they exhibit this property only at very low temperatures (usually below 25 K). Because the compound YBa$_2$Cu$_3$O$_7$ becomes superconducting at 92 K, it is known as a high-temperature superconductor; clearly the term "high temperature" is a relative one.

The structure of YBa$_2$Cu$_3$O$_7$ is that of a detect perovskite. By *defect perovskite* we mean that the metal and oxygen atoms are arranged like those in the mineral perovskite but some of the atoms are missing. The description of a compound in terms of a reference structure follows a tradition in solid-state chemistry. For example, many compounds with a 1:1 cation/anion ratio are said to adopt "the NaCl structure" or "the CsCl structure" and so on. The structures of solids are discussed in terms of their unit cell, the smallest repeating part of the structure. The unit cells are molecule-sized parallelepipeds, and when they are stacked together, the entire three-dimensional structure can be generated. Unit cells are described by the lengths of their edges and the angles at the corners. Perovskite (CaTiO$_3$) has a very symmetric unit cell: Its edges are all equal in length ($a = b = c$) and the angles α, β, and γ are all 90°. The dimensions of the unit cell define a cube. In the CaTiO$_3$ unit cell, the Ti atoms are at the corners of the cube and the oxygen atoms are situated along each of the 12 edges (Fig. 1-1). The calcium ion is at the center of the cube and is surrounded by 12 oxygen atoms. For the purpose of generalizing this structure type, the Ca atoms are said to occupy the *A* sites and the Ti atoms occupy the *B* sites. All of the structures in the perovskite family can be viewed as variations of this arrangement, and all have the general formula ABO$_3$. Because atoms in the *A*

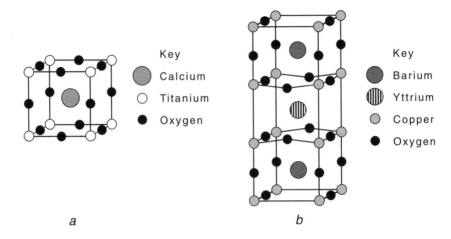

Figure 1-1
Structure of the unit cells for $CaTiO_3$ (*a*) and $YBa_2Cu_3O_7$ (*b*).

sites have higher coordination numbers than those in the *B* sites, we expect to see the larger of the two metal atoms situated in the *A* sites.

The compound ReO_3 is an example of a defect perovskite; in this material, all of the *B* sites are filled with Re atoms whereas all of the *A* sites are vacant. The compound $YBa_2Cu_3O_7$ is a relatively complicated defect perovskite. First, two kinds of atoms occupy the *B* sites, Ba^{2+} and Y^{3+}, whereas the copper atoms occupy the *A* sites; thus $YBa_2Cu_3O_7$ can be described as having the formula $A_3B_3O_7$. Second, unlike perovskite, the A/B/O ratio in $YBa_2Cu_3O_7$ is not $1:1:3$; instead, some oxygen atoms are missing. These missing oxygen atoms result in coordination numbers 8 for the Y atoms, 10 for the slightly larger Ba atoms, and either 4 or 5 for the Cu atoms. The overall structure of the unit cell is shown in Figure 1-1. Because of the aforementioned defects, the unit cell for $YBa_2Cu_3O_7$ is not cubic like $CaTiO_3$, but instead the edges are unequal in length ($a \neq b \neq c$) although the angles α, β, and γ are all still $90°$. This type of unit cell is called orthorhombic. The formula for the solid can be calculated from this drawing by counting only those parts of the atoms that lie inside the unit cell. We leave this as an exercise.

The preparation of compounds such as $YBa_2Cu_3O_7$ directly from solid precursors is an important part of inorganic chemistry. One advantage of such solid-state syntheses is that they often afford the desired product in quantitative yields with no purification needed. Solid-state reactions, however, proceed slowly at normal temperatures because atoms in crystals do not diffuse rapidly. This problem does not exist for solution-phase syntheses because of the high mobilities of solutes in solvents. The slow atomic diffusion rates for solids can be overcome by heating them to the stage where the atoms migrate quickly enough to form new chemical bonds on a reasonable time scale. The synthesis of $YBa_2Cu_3O_7$ requires several hours at temperatures of about $900°C$. Such conditions are typ-

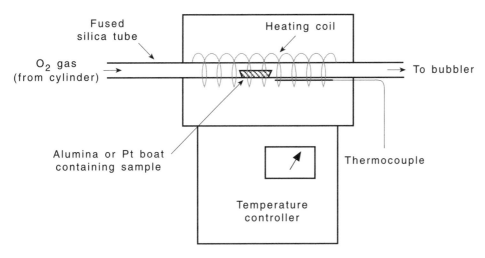

Figure 1-2
Tube furnance used for high-temperature synthesis.

ical of many solid-state inorganic syntheses. Such high temperatures require the use of some specialized equipment, however.

The most important piece of equipment for high-temperature syntheses is the tube furnace (Fig. 1-2). A tube furnace consists of a ceramic pipe wrapped with resistively heated wire. Some tube furnaces have two or more temperature zones, and many have programmable heating and cooling cycles. Not only are the temperatures important, but the rates of heating and cooling can significantly influence the purity and crystallinity of products of solid-state reactions. It is also important to select appropriate reaction vessels. Pyrex (a borosilicate glass) softens at about 800 °C, so it is common to use fused silica (a noncrystalline form of silicon dioxide), which remains solid up to 1000 °C. Another issue one must consider is the chemical compatibility of the reactor walls, because many species considered unreactive at room temperature are very reactive at elevated temperatures. In the synthesis of YBa$_2$Cu$_3$O$_7$, the reactants are placed in an alumina container to protect them from direct contact with the silica tube. Platinum containers can also be used in place of alumina containers.

An early step in this synthesis is the decomposition of BaCO$_3$ to give BaO and CO$_2$. This reaction is called *calcining*; another example of calcining is the conversion of limestone, CaCO$_3$, into lime, CaO, in the preparation of Portland cement. The high temperatures involved in the synthesis cause atomic diffusion and partial crystallization, so that the resulting product often consists of masses of microcrystalline particles, which adhere to one other. This transformation is called *sintering*.

Solids with the formula YBa$_2$Cu$_3$O$_x$ are stable for a range of oxygen stoichiometries ($6 \leq x \leq 7$). At high temperatures in the absence of O$_2$, a solid with $x = 6.5$ is formed. In the present experiment, you will heat this $x = 6.5$ compound under an O$_2$ atmosphere to give the material having $x = 7$. The value of x

strongly affects the properties of the material, including the temperature at which it becomes superconducting

$$\tfrac{1}{2}\, Y_2O_3 \; + \; 2\, BaCO_3 \; + \; 3\, CuO \; \longrightarrow \; YBa_2Cu_3O_{6.5} \; + \; 2\, CO_2$$

$$YBa_2Cu_3O_{6.5} \; + \; \tfrac{1}{4}\, O_2 \; \longrightarrow \; YBa_2Cu_3O_7$$

If the three metals Y, Ba, and Cu are not present in exactly a $1:2:3$ ratio, or if the starting materials are not well mixed, then one obtains an impure product that contains variable amounts of the nonsuperconducting phases $BaCuO_2$ and Y_2BaCuO_5.

Superconductivity

One of the most unusual properties that a material can have is superconductivity, which means that the solid conducts electricity with zero, not just small, resistance. The phenomenon was discovered by the Dutch physicist Kammerlingh Onnes in the early 1900s. He found that mercury, when cooled to within a few degrees of 0 K, conducts electricity with no resistance. The temperature at which a material becomes superconducting is termed the critical temperature, or T_c. In the years following Onnes' discovery, several oxides and alloys were found with T_c values around 20 K.

Superconducting materials are used to construct high-field magnets. It has long been known that magnetic fields are generated by passing electricity through a loop. By making the loops out of zero-resistance superconductors, large currents can be passed through them, thus creating strong magnetic fields. One advantage of superconducting magnets is that, once a current is established in the loop, the current will circulate through the loop indefinitely and no external current source is needed. Magnets for most Fourier transform nuclear magnetic resonance (FT–NMR) instruments have loops made from the superconducting alloy Nb_3Sn ($T_c = 23$ K). Aside from T_c, superconductors are classified based on their critical currents, J_c, and critical fields, H_c. Above these critical values, the superconducting properties are lost.

In 1987, two scientists, K. Alex Müller and J. Georg Bednorz, discovered a barium–lanthanum–copper oxide that becomes superconducting near 35 K. Very soon thereafter, $YBa_2Cu_3O_7$, or "1-2-3" as it is sometimes called, was discovered to have a T_c of 92 K. This announcement was followed by the synthesis of several related families of superconductors with T_c values above 100 K. The practical significance of these "high-temperature" superconductors is that they become superconducting when cooled with liquid N_2 (boiling point 77 K), a relatively inexpensive refrigerant ($\approx US\$1$ per liter). Above T_c, $YBa_2Cu_3O_7$ is a normal metal, meaning that it has a small but nonzero electrical resistivity that increases with increasing temperature. In contrast, the material $YBa_2Cu_3O_6$ is not a superconductor but instead is a semiconductor. Its resistivity decreases as the temperature is raised.

The mechanism of superconductivity remains the subject of much experimental and theoretical study. There is agreement that a mechanism exists by which the movement of one electron assists in the movement of a partner electron. These two electrons are called a Cooper pair.

One of the most dramatic properties of superconductors is their ability to exclude magnetic fields. As discussed in Experiment 12, diamagnetic compounds and magnets repel one another. Superconductors are called perfect diamagnets because the external magnetic field induces a current within the superconductor that generates an exactly equal but opposite magnetic field. This phenomenon, known as the Meissner effect, means that magnets can be levitated by a super-conductor.

Iodometric Titrations

One interesting aspect of $YBa_2Cu_3O_7$ is the unusually high formal oxidation state of the copper atoms. If the Y, Ba, and O atoms are assigned their usual charges of $+3$, $+2$, and -2, respectively, then the formula $YBa_2Cu_3O_7$ requires some of the copper atoms to have a charge of $+2$ and others to have a charge of $+3$. Interestingly, spectroscopic studies show that no copper(III) centers are present in $YBa_2Cu_3O_7$, but that instead some electrons are missing from the copper-oxygen bonds. For the purposes of the following titration, however, we will think of these missing electrons as coming solely from the copper atoms, and we will proceed as if some copper(III) centers were actually present.

In this part of the experiment, you will apply the analytical technique of iodometry to determine the oxidation state of the copper. In iodometric titra-tions, the sample oxidizes iodide (I^-) to iodine (I_2), and the amount of iodine formed is measured titrimetrically. Thus, iodometry determines the number of oxidizing equivalents in the sample.

In order to interpret the results of an iodometric titration, it is necessary to know what is counted as an oxidizing equivalent. Yttrium(III) and barium(II) are counted as zero oxidizing equivalents; that is, they are not capable of oxidiz-ing I^- to I_2. Copper(I) is also not an oxidizing equivalent: It also cannot oxidize iodide to iodine. Higher oxidation states of copper, however, are capable of oxi-dizing iodide and will be reduced to Cu^I, as shown in Eqs. 1 and 2.

$$Cu^{3+} + 2\,I^- \longrightarrow Cu^+ + I_2 \tag{1}$$

$$Cu^{2+} + I^- \longrightarrow Cu^+ + \tfrac{1}{2}\,I_2 \tag{2}$$

Thus copper(III) and copper(II) count as 2 and 1 oxidizing equivalents, respec-tively. The I_2 produced will react with excess I^- to form the brown triiodide ion, I_3^-. The triiodide ion has the same number of oxidizing equivalents as I_2 but is more soluble in water. From Eqs. 1 and 2, you should be able to predict the number of equivalents of I^- that will be oxidized by one equivalent of $YBa_2Cu_3O_7$.

In order to measure the amount of I_2 produced, and hence the number of oxidizing equivalents in the sample, the iodine generated from Eqs. 1 and 2 is titrated with sodium thiosulfate, a reducing agent sometimes referred to as "hypo" (Eq. 3).

$$\tfrac{1}{2}\,I_2 + S_2O_3^{2-} \longrightarrow I^- + \tfrac{1}{2}\,S_4O_6^{2-} \tag{3}$$

The endpoint is reached when the iodine is consumed; to aid in determining the endpoint, starch is added toward the end of the titration to form the intensely blue starch–triiodide complex. Note that each thiosulfate ion furnishes one reducing equivalent.

In order to determine the concentration of the standard $S_2O_3^{2-}$ solution, it will be used to titrate a known amount of I_2, which will be generated through the comproportionation reaction:

$$5\,I^- + IO_3^- + 6\,H^+ \longrightarrow 3\,I_2 + 3\,H_2O \tag{4}$$

The KIO_3 can be purchased as a ready-made standard solution. The other reactants are present in excess.

EXPERIMENTAL PROCEDURE

Note: Avoid the use of metal spatulas throughout this experiment; metal impurities diminish the superconducting properties of the product. The reactants used in this experiment are mildly toxic, and the dusts should not be inhaled. Avoid eye or skin contact and wash your hands after handling the powders.

Yttrium Barium Copper Oxide, YBa₂Cu₃O₇

Weigh out 225.8 mg (1 mmol) of Y_2O_3, 789.4 mg (4 mmol) of $BaCO_3$, and 477.4 mg (6 mmol) of CuO. Note that these weights correspond to a $1:2:3$ molar ratio of metals; try to make sure that the actual masses are within a few tenths of a milligram of the masses given here. In a hood, combine the powders in a clean agate mortar and grind the mixture with an agate pestle until the powders are thoroughly mixed (about 15 min). The final powder should be of a uniform gray color with no lumps. Transfer the gray powder into a preweighed platinum or alumina boat and record the total weight of the starting reagents. Place the boat near the center of the tube in the furnace. Position the tube in the furnace next to a thermocouple (Fig. 1-2). Because the furnace temperature is controlled by the thermocouple, it is important to position the thermocouple probe near the reaction boat (but on the outside of the tube). Fit the ends of the tube with gas adapters, connecting one adapter to an oxygen cylinder and the other end to an oil bubbler so that the gas flow can be monitored. Before turning on the furnace, adjust the oxygen flow to 1–2 bubbles per second. Then heat the sample to 970 °C for 2–3 h.

Turn off the power to the furnace (the temperature controller should be left on), leaving the oxygen flowing. Allow the temperature to reach 700 °C before opening the furnace to accelerate the cooling. Once the sample has cooled to 200 °C, turn off the oxygen, and disassemble the apparatus. When the tube is cool enough to handle, remove it from the furnace. The sample should be black, and the particles should be sintered together. Weigh the boat with the sample still in it, and regrind the sample to a fine powder; the particles should be black and lustrous. *Here is a good stopping point for the first day's work.*

Meissner Effect and Electrical Resistance

Safety note: Liquid N$_2$ can cause severe frostbite. Immediately remove clothing that becomes saturated with liquid nitrogen, because the liquid held by the fabric may freeze the skin underneath and cause severe burns.

The presence of the superconducting phase can be tested by the ability of your sample to repel a magnetic field. Use a pellet press to make a small pellet out of some of your black powder. See your instructor for use of the pellet press; if a press is not available, follow the procedure described in the next paragraph. With plastic forceps, place the pellet in a cut-off styrofoam cup. With the same plastic forceps, place a small magnet on top of the pellet, and then carefully pour enough liquid N$_2$ into the cup to cover the pellet. When the pellet cools sufficiently, the magnet should "levitate." If you touch the magnet with plastic forceps, you should be able to make it spin.

If a pellet press is not available, place some liquid N$_2$ in a concave surface (e.g., the bottom of an empty inverted aluminum can). Position a small magnetic stir bar in the center of the concave surface, and add some of the black powder to the liquid N$_2$. Once the temperature of the superconductor drops below T_c, the powder should migrate away from the magnet. If the powder fails to become superconducting, reheat it to 970 °C overnight under a flow of oxygen and then test the Meissner effect again.

If the necessary apparatus is available, measure the electrical resistance of a pressed pellet of the 1-2-3 compound as it is cooled from room temperature to liquid N$_2$ temperature. See your instructor or the book by Ellis et al. listed at the end of this experiment for a description of this procedure.

Iodometric Titration of YBa$_2$Cu$_3$O$_7$

Of the following four solutions, the sodium thiosulfate and starch solutions must be prepared the same day as the titration; the other two can be prepared in advance if desired.

Prepare a sodium thiosulfate solution from 3.3 g of Na$_2$S$_2$O$_3$·5 H$_2$O and 250 mL of H$_2$O.

Prepare a starch indicator by adding 0.2 g of soluble starch to 100 mL of H$_2$O, heating the solution in a water bath until the starch dissolves and the solution is clear.

Prepare a 3.5 *M* HCl solution by slowly adding 30 mL of concentrated 12 *M* HCl to 60 mL of distilled H$_2$O (*Caution:* heat evolved!) and then diluting the mixture with H$_2$O to a final volume of 100 mL.

Prepare a 10% potassium iodide solution by dissolving 10 g of KI in 100 mL of H$_2$O.

For the following titration, have a squirt bottle of deionized H$_2$O on hand to wash titrant off the tip of the buret, the sides of the Erlenmeyer, and the nitrogen pipet.

Using standard techniques, charge a 50-mL buret with the sodium thiosulfate solution to be standardized. Take a volumetric pipet and rinse it with a few mL of the KIO$_3$ solution. Discard the rinse, and then measure out 10.00 mL of the KIO$_3$ solution (the molarity is one-sixth of the normality printed on the bottle, because each mole of IO_3^- furnishes six oxidizing equivalents according to Eq. 4). Charge a 125-mL Erlenmeyer flask with the 10.00 mL of the KIO_3 solution, and then add 10 mL of distilled H_2O, 10 mL of 10% KI solution, and 10 mL of 3.5 M HCl. The solution should turn brown. Note the initial volume of the sodium thiosulfate solution in the buret and begin the titration by adding 21 mL of it to the Erlenmeyer flask. Now add 3 mL of starch solution and titrate (1 drop every 4 s) until the blue color disappears.

Calculate the concentration of the sodium thiosulfate solution and repeat the titration to ensure a reproducible value.

Weigh out about 0.11 g of the $YBa_2Cu_3O_7$ material (note the exact weight) into a clean 125-mL Erlenmeyer flask. Add 15 mL of 10% potassium iodide solution; the material will not dissolve. Remove dissolved O_2 by bubbling N_2 through the solution vigorously for 10 min. To do this, attach a Pasteur pipet to a hose, connect the hose to a nitrogen tank, and clamp the pipet near the mouth of the Erlenmeyer flask. Make sure the tip of the pipet is immersed in the solution. Throughout the rest of the titration, leave the nitrogen bubbling gently through the solution. Allow the redox process to take place by adding 6 mL of the 3.5 M HCl solution over about 5 min. The solution should turn brown due to the generation of the triiodide ion. Swirl the solution for about 10 min to allow the solid to react; the result will be a brown solution and a gray precipitate. Note the initial volume of the standardized sodium thiosulfate titrant and begin the titration by adding 10 mL of it to the brown solution. The color should become much paler. Now add 3 mL of the starch solution; a blue color should form. Continue titrating slowly (1 drop every 5 s). When the blue color becomes paler, titrate at one-half that speed. The endpoint is reached when the blue color disappears, leaving a white, turbid mixture that remains white for 30 s (false endpoints may occur). Record the final volume of titrant in the buret.

REPORT

Include the following:
1. Weight change when the solids are heated in the boat. Estimate the value of x in the formula $YBa_2Cu_3O_x$ from the weight change.
2. Experimental and theoretical number of equivalents of I^- oxidized by 1 g of your sample.
3. Average oxidation state of the copper atoms, and the value of x in the formula $YBa_2Cu_3O_x$, from the results of the iodometric titration.

PROBLEMS

1. The compound $YBa_2Cu_3O_7$ is black, whereas Y_2BaCuO_5, which is an electrical insulator, is green. Account for this difference. In your answer assign oxidation states to all of the metals in these materials.
2. Why are high temperatures required in the synthesis of $YBa_2Cu_3O_7$?
3. Calculate the unit cell dimensions a, b, and c for $YBa_2Cu_3O_7$ from the indexed X-ray powder pattern provided in Appendix 10. Explain why the crystals

are nearly tetragonal in terms of the atomic structure of the compound. The following formula is useful:

$$\sin^2\theta = \frac{\lambda^2}{4}\left(\frac{h^2}{a^2} + \frac{k^2}{b^2} + \frac{l^2}{c^2}\right)$$

The three numbers, *hkl*, called Miller indices, indicate the direction of the scattering plane (see Experiment 2 for an explanation of diffraction patterns). The Miller indices can be read from the diffractogram provided. For simple reflections of the type *h*00, 0*k*0, and 00*l*, the value of *h*, *k*, or *l* corresponds to *n* in Bragg's law (see Experiment 2).

4. The compound YBa$_2$Cu$_3$O$_7$ is a defect perovskite. What would be the "nondefect" formula for the YBa$_2$Cu$_3$O$_x$ compound?
5. Why is N$_2$ bubbled through the solution when titrating the triiodide ions generated from YBa$_2$Cu$_3$O$_7$?

INDEPENDENT STUDIES

A. Prepare the "green phase" Y$_2$BaCuO$_5$ by the reaction of the metal oxides in the metal atom ratio 2:1:1 and use it to make YBa$_2$Cu$_3$O$_7$. (Cogdell, C. D.; Wayment, D. G.; Casadonte, D. J. Jr.; Kubat-Martin, K. A. *J. Chem. Educ.* **1995**, *72*, 840.)

B. Use a ball mill to mix the three starting materials before making YBa$_2$Cu$_3$O$_7$ in the tube furnace. This preparation has the advantage that an oxygen gas cylinder is not needed. (Garbauskas, M. F.; Arendt, R. H.; Kasper, J. S. *Inorg. Chem.* **1987**, *26*, 3191.)

C. Record the X-ray powder diffraction pattern of YBa$_2$Cu$_3$O$_x$ as *x* varies from 6 to 7 and determine how the lattice constants change.

D. Use a four-point probe and a thermocouple to measure the electrical resistance of a pressed pellet of YBa$_2$Cu$_3$O$_7$ between liquid N$_2$ and room temperature.

E. Prepare a different high-temperature superconductor such as La$_{2-x}$Sr$_x$CuO$_4$. (Krajewski, J. J. *Inorg. Synth.* **1995**, *30*, 192.)

F. Use a potentiometric titration to determine the oxidation state of the copper atoms in YBa$_2$Cu$_3$O$_7$. (Phinyocheep, P.; Tang, I. M. *J. Chem. Educ.* **1994**, *71*, A115.)

REFERENCES

High-Temperature Superconductors
Bednorz, J. G.; Müller, K. A. *Z. Phys. B* **1986**, *64*, 189. The paper that sparked the search for high T_c materials; the authors were awarded the Nobel prize for these results.
Edwards, P. P.; Rao, C. N. R. *Chem. Brit.* **1994**, 722. A review of recent advances in the field.
Harris, D. C. *Quantitative Chemical Analysis*, 4th ed., Freeman: New York, 1995, pp 797–800. Titrimetric analysis of YBa$_2$Cu$_3$O$_7$.
Harris, D. C.; Hills, M. E.; Hewston, T. A. *J. Chem. Educ.* **1987**, *64*, 847. Iodometric analysis of YBa$_2$Cu$_3$O$_7$.

Narlikar, A., Ed. *Studies of High Temperature Superconductors*, Nova Science: New York, 1988–1991; Vols. 1–8.

Schneemeyer, L. F.; Waszczak, J. V.; van Dover, R. B. *Inorg. Synth.* **1995**, *30*, 210. The preparation of single crystals of $YBa_2Cu_3O_7$.

Simon, R.; Smith, A. *Superconductors. Conquering Technology's New Frontier*, Plenum: New York, 1988. A very readable layman's guide to superconductors.

Sleight, A. W. *Science* **1988**, *242*, 1519. An inorganic chemist's view of high T_c materials.

Steinfink, H.; Swinnea, J. S.; Sui, Z. T.; Hsu, H. M.; Goodenough, J. B. *J. Am. Chem. Soc.* **1987**, *109*, 3348. Structure of $YBa_2Cu_3O_7$.

Wu, M. K.; Ashburn, J. R.; Torng, C. J.; Hor, P. H.; Meng, R. L.; Gao, L.; Haung, Z. J.; Wang, Y. Q.; Chu, C. W. *Phys. Rev. Lett.* **1987**, *58*, 908. Report of the first high T_c superconductor.

General References to Solid-State Chemistry

http://www.shef.ac.uk/chemistry/chemdex/materials.html. A useful list of educational software, chemical and instrument suppliers, and databases for materials science.

Adams, D. M. *Inorganic Solids*, Wiley: New York, 1974.

Bruce, D. W.; O'Hare, D. *Inorganic Materials*, 2nd ed., Wiley: New York, 1997.

Cheetham, A. K.; Day, P., Eds. *Solid State Chemistry: Techniques*, Oxford University: Oxford, 1987. Excellent survey of methods used in solid-state inorganic synthesis.

Cheetham, A. K.; Day, P., Eds. *Solid State Chemistry: Compounds*, Oxford University: Oxford, UK, 1992.

Ellis, A. B.; Geselbracht, M. J.; Johnson, B. J.; Lisensky, G. C.; Robinson, W. R. *Teaching General Chemistry: A Materials Science Companion*, American Chemical Society: Washington DC, 1993, p 431. A description of the four-probe method to measure electrical conductivity of superconductors.

Kittel, C. *Introduction to Solid State Physics*, 6th ed., Wiley: New York, 1986.

Murphy, D. W.; Interrante, L. V., Eds. *Inorganic Syntheses,* Vol. 30, Wiley: New York, 1995. This entire issue is dedicated to the synthesis of solid-state materials.

Smart, L.; Moore, E. *Solid State Chemistry, An Introduction*, Chapman and Hall: London, 1992.

Wold, A.; Dwight, K. *Solid State Chemistry*, Chapman and Hall: London, 1993.

Wells, A. F. *Structural Inorganic Chemistry*, 5th ed., Oxford University: Oxford, UK, 1984.

The Layered Solids $VO(PO_4)(H_2O)_2$ and $VO(HPO_4)(H_2O)_{0.5}$

Note: This experiment requires 4 hours spread over two laboratory periods.

Many inorganic solids consist of densely packed structures, while others have more spacious lattices, sometimes with molecule-size channels and voids (see Experiment 3). Such open-framework structures are of interest because they allow molecules, sometimes referred to as "guests," to travel through the interior of the solid, or "host." In zeolites, small molecules move through rigid channels. In this experiment, you will prepare a different kind of open framework structure, one that consists of layers of atoms. Such layered structures are important because a greater variety of guest atoms and molecules can enter the lattice and can then interact with the atoms in the interior of the framework.

The most famous example of a layered material is graphite. In graphite, the layers consist of three-coordinate carbon atoms arranged in six-membered rings; the six-membered rings are interconnected to give a hexagonal pattern (Fig. 2-1).

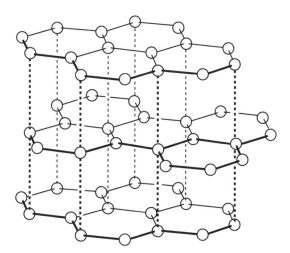

Figure 2-1
Structure of graphite. The dotted lines indicate the relative positions of atoms in adjacent layers.

The bonding within the graphite layers is very strong and consists of C–C bonds like those in benzene. The bonding between the layers, however, is weaker and consists of van der Waals interactions. Consequently, the layers slide readily over one another giving graphite a slippery feel and making it a useful lubricant. Furthermore, the weak interlayer bonding allows various reagents to enter the spaces between the layers.

Although both zeolites and layered materials serve as hosts for molecular guest molecules, there is an important difference between these host materials. In zeolites, the channels and voids are preformed and have well-defined sizes, whereas in two-dimensional materials the voids are formed by prying the layers apart so that the sizes of the voids can in principle be increased without limit. In layered materials, the interactions between the guests and the layers must be sufficient to compensate for the disruption of the van der Waals interactions that exist between the layers. For example, potassium reacts with graphite to afford solids with formulas like KC_{24} and KC_8. This reaction, called *intercalation*, involves widening the gap between the carbon sheets and inserting the potassium atoms into the gap. The intercalation of potassium is favored because electrons are transferred from the alkali metal to the carbon sheets: The resulting potassium cations are electrostatically attracted to the negatively charged carbon layers. Unless atoms and molecules can donate electrons to or accept electrons from the carbon sheets, they will not intercalate. Intercalation can be reversible; for example, treatment of KC_8 with Br_2 gives back graphite and generates KBr.

The compound $VO(PO_4)(H_2O)_2$ also has a layered structure. For $VO(PO_4)(H_2O)_2$, the layers consist of $VO(PO_4)(H_2O)$ units in which each vanadyl (VO) center is linked to four separate phosphate groups and one water ligand. A top view of a sheet is shown in Figure 2-2.

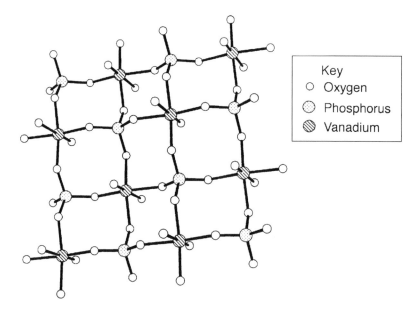

Figure 2-2
Fragment of $VO(PO_4)(H_2O)_2$ structure showing part of one of the layers. Hydrogen atoms and one-half of the water molecules are not shown.

Two types of water molecules are present in $VO(PO_4)(H_2O)_2$. One of the water molecules per formula unit is coordinated to vanadium, trans to the V=O group. The second water molecule in $VO(PO_4)(H_2O)_2$ is situated between the layers; both water molecules hold the layers together by hydrogen bonding to the oxide framework.

In $VO(PO_4)(H_2O)_2$, the vanadium atoms are in their highest (+5) oxidation state. This compound is prepared by the reaction of V_2O_5 with phosphoric acid, H_3PO_4, in the presence of a small amount of nitric acid:

$$V_2O_5 + 2\,H_3PO_4 + H_2O \longrightarrow 2\,VO(PO_4)(H_2O)_2$$

Vanadium centers in the +5 oxidation state are potent oxidants, and vanadium(V) oxides are good oxidation catalysts. For example, vanadium phosphates are used in industry as catalysts for the oxidation of butane to maleic anhydride:

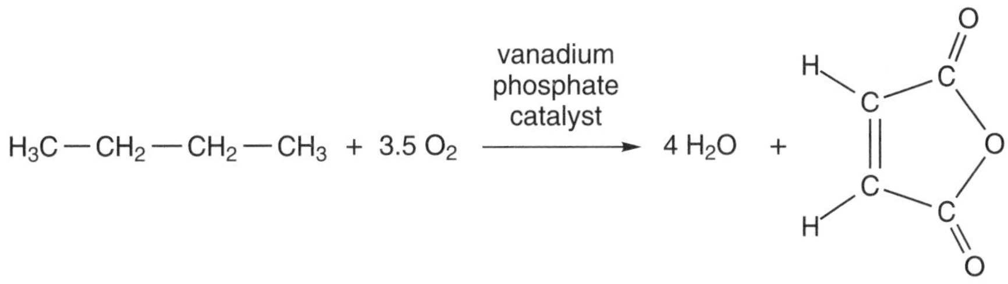

Solid catalysts work best if they have high surface areas; this maximizes contact of the catalyst with the reactants. Usually, high surface areas must be generated by grinding the solid catalyst to a fine powder. In the present case, however, grinding is not as crucial: The layered structure of $VO(PO_4)(H_2O)_2$ gives it a large effective surface area because the reactants can enter between the layers.

In the present experiment, you will use $VO(PO_4)(H_2O)_2$ to oxidize the alcohol 2-butanol to the ketone 2-butanone. Because this oxidation does not involve oxygen atom transfer, the vanadium phosphate framework remains essentially intact throughout the cycle. The balanced equation is

$$2\,VO(PO_4)(H_2O)_2 + CH_3CH_2CH(OH)CH_3$$
$$\longrightarrow 2\,VO(HPO_4)(H_2O)_{0.5} + CH_3CH_2COCH_3 + 3\,H_2O$$

Some aspects of the mechanism of the oxidation process can be inferred from the coordination chemistry of $VO(PO_4)(H_2O)_2$. For example, it is thought that the alcohol binds directly to the oxidizing vanadium center by replacing the coordinated water ligand. This conclusion is based on studies showing that the water ligand can be replaced by other Lewis bases. For example, $VO(PO_4)(H_2O)_2$ reacts with pyridine to give the adduct $VO(PO_4)(NC_5H_5)$.

The oxidation of the alcohol changes the oxidation state of the vanadium from +5 to +4. The redox reaction is marked by the conversion of the yellow vanadium(V) compound to a blue vanadium(IV) product. The d^1 configuration of vanadium(IV) means that the product, $VO(HPO_4)(H_2O)_{0.5}$, is paramagnetic.

The blue color arises from an electronic transition of the d electron on the vanadium(IV) center.

Powder X-Ray Diffraction

A very powerful method for the characterization of inorganic solids is powder X-ray diffraction. In this technique, a beam of X-rays is directed at a powdered sample of the compound to be studied. The X-ray beam is scattered by the atoms in the sample and is detected. If the sample is crystalline, constructive and destructive interference cause the X-rays to be scattered only in certain directions. The intensities of the scattered X-rays are determined as a function of the angle between the incident beam and the scattered ray. The plot of the intensity of the scattered X-rays versus angle is called a diffraction pattern, which is different for different substances. One can often match the observed diffraction pattern with that of a known compound, and in this way one can often quickly identify what the solid sample is made of. Powder X-ray diffraction patterns can also be used to determine the structures of new compounds.

Crystalline solids give X-ray diffraction patterns that contain sharp peaks. In contrast, amorphous or noncrystalline substances (such as glass or liquid water) scatter X-rays in all directions and give X-ray diffraction patterns that contain no sharp peaks. What is a crystalline solid? Crystalline solids are constructed from identical blocks of a basic structure that are repeated in space. The shape of each block (which is called a "unit cell") is defined by the lengths of its edges and the angles between them. The lengths of the edges in the x, y, and z directions are designated a, b, and c, respectively, and it is necessary to specify only three angles, called α, β, and γ, to define all the angles between the edges. The unit cell can be cubic ($a = b = c$ and all angles $= 90°$), brick-shaped (edges different lengths but all angles $= 90°$), or parallelepiped-shaped (all lengths and angles different).

The interpretation of X-ray powder diffraction patterns relies on Bragg's law:

$$n\lambda \ = \ 2\,d\sin\theta$$

where n is an integer, λ is the wavelength of the X-rays, d is the spacing between identical planes of atoms in the solid, and θ is the incidence angle (see Fig. 2-3). The number and types of atoms in each plane determine the *intensity* of the diffraction peaks; the spacing between the planes (d) determines the *angle* between the incident and the scattered beams (which is equal to twice the incident angle of θ). For historical reasons, the angle between the incident and scattered beams (2θ) is used as the x axis of diffraction patterns.

Bragg's law states the conditions under which the identical planes of atoms will scatter X-rays constructively. Constructive interference occurs only when the X-rays scattered from different planes follow paths whose path lengths differ by an integer multiple of the X-ray wavelength.

As described previously, every peak in an X-ray diffraction pattern corresponds to scattering from a specific set of planes separated by a distance d. The planes are all parallel with one another, and their orientation in space can be defined by three numbers that describe, in order, how the planes intersect the x, y,

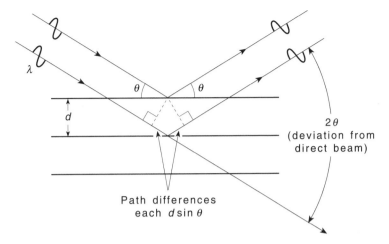

Figure 2-3
Reflection of X-rays by an idealized crystal lattice. The Bragg law states that, for diffraction to occur, the path difference $2d \sin \theta$ must be equal to an integer multiple of the X-ray wavelength λ.

and z edges of the unit cell. The three numbers, which are called Miller indices, are conventionally written with no spaces between them: For example, 001 means that the three Miller indices are 0, 0, and 1. Although a detailed description of Miller indices is beyond the scope of the present discussion, it is useful to know that the Miller indices are related to the incidence angle θ and the interplanar spacing d. For a crystal in which $\alpha = \beta = \gamma = 90°$, the following equation holds:

$$\frac{2 \sin \theta}{\lambda} = \frac{1}{d} = \sqrt{\left(\frac{h}{a}\right)^2 + \left(\frac{k}{b}\right)^2 + \left(\frac{l}{c}\right)^2}$$

where h is the first Miller index, k is the second, and l is the third.

In this experiment, you will analyze the X-ray powder diffraction pattern of $VO(PO_4)(H_2O)_2$. Even though your sample is a powder and may not look crystalline to your eye, under a microscope the individual powder particles have characteristic crystalline shapes, and the X-ray powder diffraction pattern will look just like one obtained from a crystal of larger size. The diffraction patterns of layered materials such as $VO(PO_4)(H_2O)_2$ are instructive because they are often fairly simple. Because the density of atoms in the sheets is high for layered compounds, the reflections from these planes of atoms are often quite intense. In $VO(PO_4)(H_2O)_2$, which consists of layers of atoms perpendicular to the c axis of the crystal, the most intense reflection in the diffraction pattern, the 001 reflection, is found at $2\theta = 11.88°$ when copper $K\alpha$ radiation ($\lambda = 1.54$ Å) is used. This reflection satisfies Bragg's law for $n = 1$. Also diagnostic of layered materials is the series of reflections corresponding to the other integer values of n. These are called higher order reflections.

Changes in the interlayer distance d can be easily detected by X-ray diffrac-

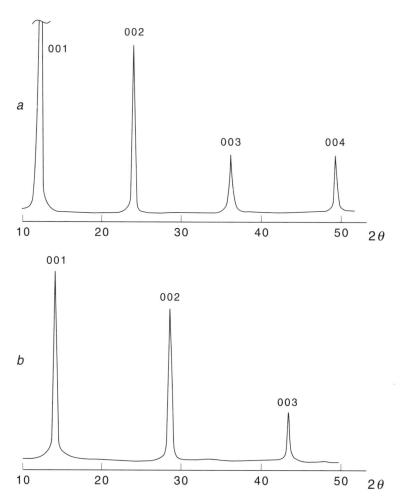

Figure 2-4
Powder X-ray diffraction patterns for $VO(PO_4)(H_2O)_2$ (a) and $VO(PO_4)(H_2O)$ (b) using copper $K\alpha$ radiation ($\lambda = 1.54$ Å).

tion. Focusing only on the $n = 1$ case, the reflection will occur at a larger angle if d decreases. For example, heating $VO(PO_4)(H_2O)_2$ gives the monohydrate $VO(PO_4)(H_2O)$. The powder pattern of the monohydrate shows that the layer structure is retained, but the 001 reflection now appears at $2\theta = 14.18°$ (Fig. 2-4). The increase in the 2θ value (from $11.88°$ for the dihydrate) shows that the layers have been brought closer together; this happens because heating drives out the water molecules between the layers. The higher order reflections also shift to larger angles. Conversely, intercalation of large molecules such as pyridine causes the layers to move farther apart and the 2θ values to become smaller.

EXPERIMENTAL PROCEDURE

Safety note: Vanadium pentoxide dust is toxic and potentially carcinogenic. It should be handled in a well-ventilated hood.

Vanadyl Phosphate Dihydrate, VO(PO$_4$)(H$_2$O)$_2$

Vanadium(V) compounds are strong oxidants. This procedure will fail if the reaction mixture is exposed to hydrocarbons such as trace residues of organic (wash) solvents or stopcock grease. The V$_2$O$_5$ itself should be stored in the absence of organic vapors and freshly ground before use. Do not grease the joints of the glassware used in this experiment.

In a hood, grind about 1.2 g of V$_2$O$_5$ using a mortar and pestle. Place 0.96 g (5.28 mmol) of this powder in a 50-mL round-bottom flask together with a clean Teflon-coated stir bar. Then add in order 5.3 mL of concentrated H$_3$PO$_4$, 11 mL of H$_2$O, and 3 drops of concentrated HNO$_3$ (*Caution: very corrosive!*). Attach a reflux condenser and start the water flowing. Next, heat the slurry to reflux for 2 h or longer. (Longer reaction times increase the crystallinity of the product.) After 2 h allow the bright yellow slurry to cool to room temperature. Filter the product on a sintered glass frit in air, and then wash the yellow powder with four 5-mL portions of water followed by 5–10 mL of reagent grade acetone. Dry the solid in air and save some so that you can take its infrared (IR) spectrum. A green tint indicates that trace amounts of vanadium(IV) impurities are present. These impurities should not adversely affect the second part of the experiment.

Vanadyl Monohydrogenphosphate Hemihydrate, VO(HPO$_4$)(H$_2$O)$_{0.5}$

Place 0.5 g (1.71 mmol) of VO(PO$_4$)(H$_2$O)$_2$ and 10 mL of 2-butanol in a 25-mL round-bottom flask containing a Teflon-coated magnetic stir bar. Attach a reflux condenser, turn on the water, and heat the mixture to reflux. *Here is a good stopping point for the first day's work.* You should leave the mixture refluxing until the next day.

After the solution has refluxed for 21–24 h, let the mixture cool to room temperature and filter off the blue solid. Dry the sample in air. Test the filtrate for the presence of 2-butanone using 2,4-dinitrophenylhydrazine reagent (DNP), if available. Run a control experiment by adding DNP to some fresh 2-butanol.

Take IR spectra of both VO(PO$_4$)(H$_2$O)$_2$ and VO(HPO$_4$)(H$_2$O)$_{0.5}$ (see Experiment 19 for the preparation of a Nujol mull).

Powder X-Ray Diffraction Pattern of VO(PO$_4$)(H$_2$O)$_2$

If a powder diffractometer is available, measure the powder diffraction pattern of VO(PO$_4$)(H$_2$O)$_2$. Consult your instructor on the operation of the powder X-ray diffractometer. These are expensive instruments, and there are hazards associated with X-radiation and with the large power supply that drives the X-ray generator. Record the wavelength of X-rays used. If an X-ray powder diffractometer is not available, consult Figure 2-4.

REPORT

Include the following:
1. Yields of $VO(PO_4)(H_2O)_2$ and $VO(HPO_4)(H_2O)_{0.5}$.
2. Infrared spectra of your products.
3. Indexed powder X-ray diffraction pattern for $VO(PO_4)(H_2O)_2$. Calculate the length of the c axis for the compounds prepared in this experiment.

PROBLEMS

1. What is the structure of $VO(HPO_4)(H_2O)_{0.5}$ and what is the charge of the "HPO_4" unit?
2. What products would you expect from the following reaction? (Unless indicated otherwise, the oxygen atoms are ^{16}O.)

$$2\ VO(PO_4)(H_2O)_2\ +\ CH_3CH(^{18}OH)CH_2CH_3\ \longrightarrow\ ?$$

3. Which coordination geometries are typical of vanadyl complexes?
4. Explain qualitatively the colors of the V^{4+} and V^{5+} compounds prepared in this project.
5. The c axis of the unit cell for $VO(PO_4)$(pyridine) is 9.59 Å long (this is also the distance between the layers in the structure). Calculate 2θ for the 001 peak for Cu $K\alpha$ X-rays ($\lambda = 1.54$ Å). (See Johnson, J. W.; Jacobson, A. J.; Brody, J. F.; Rich, S. M. *Inorg. Chem.* **1982**, *21*, 3820.)

INDEPENDENT STUDIES

A. Prepare the layered vanadyl material $VO(C_6H_{13}PO_3)(H_2O)_2$ using hexylphosphonic acid, $C_6H_{13}PO_3H_2$, in place of phosphoric acid. (Brody, J. F.; Johnson, J. W. *Inorg. Synth.* **1995**, *30*, 241.)
B. Determine the magnetic moment of $VO(HPO_4)(H_2O)_{0.5}$ (see Experiment 12).
C. Prepare $VO(PO_4)$(py) by reaction of $VO(PO_4)(H_2O)_2$ with pyridine. See if you can find other substituted pyridines that might push the V–O–P layers farther apart.
D. Monitor the conversion of the 2-butanol to 2-butanone using gas chromatography.
E. Use thermogravimetric analysis (TGA) to study the dehydration of $VO(PO_4)(H_2O)_2$.
F. Prepare the molecular vanadium phosphate $[C_8H_{11}NH]_4[(VO)_4(P_2O_7)-(OCH_3)_4]$ from $VO(SO_4)\cdot4\ H_2O$, trimethylpyridine, and pyrophosphoric acid. (Herron, N.; Thorn, D. L.; Harlow, R. L.; Coulston, G. W. *J. Am. Chem. Soc.* **1997**, *119*, 7149.)

REFERENCES

Vanadium Phosphates

Centi, G.; Trifiro, F.; Ebner, J. R.; Franchetti, V. M. *Chem. Rev.* **1988**, *88*, 55. A review of metal phosphates.

Ellison, I. J.; Hutchings, G. J.; Sananes, M. T.; Volta, J.-C. *J. Chem. Soc., Chem. Commun.* **1994**, 1093. Effect of alcohol chain length on the structure of vanadium(IV) phosphate reduction product.

Johnson, J. W.; Jacobson, A. J.; Brody, J. F.; Lewandowski, J. T. *Inorg. Chem.* **1984**, *23*, 3842. Synthesis of the layered organophosphonates $VO(RPO_3)(H_2O)_2$.

Johnson, J. W.; Jacobson, A. J.; Butler, W. M.; Rosenthal, S. E.; Brody, J. F.; Lewandowski, J. T. *J. Am. Chem. Soc.* **1989**, *111*, 381. Selective binding of alcohols by $VO(RPO_3)$.

Johnson, J. W.; Johnston, D. C.; Jacobson, A. J.; Brody, J. F. *J. Am. Chem. Soc.* **1984**, *106*, 8123. Characterization of $VO(HPO_4)(H_2O)_{0.5}$ and its conversion to $(VO)_2P_2O_7$.

R'kha, C.; Vandenboore, M. T.; Livage, J.; Prost, R.; Huard, E. *J. Solid State Chem.* **1986**, *63*, 202. Contains the powder X-ray diffraction pattern for $VO(PO_4)(H_2O)_2$.

Soghomonian, V.; Qin, C.; Haushalter, R. C. *Science* **1993**, *259*, 1596. Hydrothermal synthesis, structure, and magnetism of a chiral vanadium phosphate.

Layered Compounds

Mallouk, T. E.; Gavin, J. A. *Acc. Chem. Res.* **1998**, *31*, 209. Molecular recognition in lamellar solids and thin films.

Schöllhorn, R. *Angew. Chem., Int. Ed. Engl.* **1980**, *19*, 983. Review of intercalation compounds.

Thompson, M. E. *Chem. Mater.* **1994**, *6*, 1168. Review of layered metal phosphonates.

Wells, A. F. *Structural Inorganic Chemistry*, 5th ed., Oxford University: Oxford, UK, 1984.

Whittingham, M. S.; Jacobson, A. J. *Intercalation Chemistry*, Academic: New York, 1982.

X-Ray Diffraction

http://www.iumsc.indiana.edu. Includes links to software, tutorials, and research centers for X-ray diffraction.

http://www.shef.ac.uk/chemistry/chemdex/crystallography.html. An extensive list of crystallography-related web pages.

Adams, D. M. *Inorganic Solids*, Wiley: New York, 1974.

Butera, R. A.; Waldeck, D. H. *J. Chem. Educ.* **1997**, *74*, 115. X-ray diffraction of alloys.

Cheetham, A. K.; Day, P., Eds. *Solid State Chemistry: Techniques*, Oxford University: Oxford, UK, 1987.

Cullity, B. D. *Elements of X-ray Diffraction*, 2nd ed., Addison-Wesley: Reading, MA, 1978.

Klug, H. P.; Alexander, L. E. *X-ray Diffraction Procedures for Polycrystalline and Amorphous Materials*, 2nd ed., Wiley: New York, 1974.

Wold, A.; Dwight, K. *Solid State Chemistry*, Chapman and Hall: London, 1993. Techniques for the synthesis and characterization of inorganic solids.

The Molecular Sieve Zeolite-X

Note: This experiment requires 4–6 hours for Part A and 2 hours for Part B.

In most crystalline solids, the atoms are densely packed. In diamond, for example, the carbon atoms form a tightly knit three-dimensional network that has no voids large enough to accommodate other atoms or molecules. In some crystalline solids, however, the atoms are connected into more open networks interspersed with cavities or tunnels that are large enough to hold small molecules. Such "microporous" solids have many commercial uses: For example, they are often found in the desiccant packets used to protect objects from humidity, they are the key components in certain kinds of water softeners (different from those based on ion exchange resins discussed in Experiment 10), and they are important in the petroleum industry as catalysts for converting heavy oil fractions into gasoline.

The present experiment deals with one of the most important families of microporous solids, the molecular sieves. These solids have tunnels that pass entirely through the crystal: The diameters of the tunnels are typically between 4 and 10 Å. Some small molecules can enter these tunnels, whereas molecules that are larger than the diameter of the tunnels cannot. Consequently, these microporous solids are called "molecular sieves" to emphasize the fact that only molecules below a certain size can pass through their interiors.

The most commonly used molecular sieves are aluminosilicates, a family of solids that includes both naturally occurring minerals and artificially synthesized materials. As you might guess from the name, aluminosilicates contain silicon, aluminum, and oxygen, but they often contain other elements, especially alkali metals. Each silicon and aluminum atom is surrounded by four oxygen atoms in a tetrahedral arrangement. The four oxygen atoms in each tetrahedron often bridge to other Si or Al centers so that the tetrahedra are linked together by sharing vertexes. The tetrahedra can link up to form rings, chains, sheets, or three-dimensional networks:

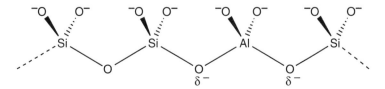

37

In pure silicates such as quartz (a crystalline form of SiO_2), the silicate tetrahedra form a relatively densely packed (and thus nonporous) three-dimensional network. If one replaces some of the Si^{4+} atoms in a silicate lattice by Al^{3+}, it is also necessary to add a unipositive cation such as sodium to maintain charge balance. These unipositive cations (which are coordinated to water molecules and thus occupy a considerable volume) can only be accommodated by forming cavities in the silicate lattice. If the cavities are sufficiently large and are suitably arranged, they can form tunnels that run entirely through the solid. Such microporous aluminosilicates are called zeolites.

Zeolites are prepared by heating a mixture of silica (SiO_2), alumina (Al_2O_3), and a hydroxide salt in water. Occasionally, other starting materials such as sodium silicate (Na_2SiO_3) and sodium aluminate ($NaAlO_2$) are used in place of the silica and alumina. Different zeolites are prepared by varying the relative ratios of the ingredients, by changing the order of addition of reagents, by employing different hydroxide salts, or by changing the temperature or pH of the reaction.

The goal of this experiment is to synthesize an X-type zeolite. This zeolite has the same structure as faujasite, a naturally occurring but rare mineral. The general formula of synthetic X-type zeolites is $A_nSi_{24-n}Al_nO_{48} \cdot m\,H_2O$, where A is a univalent ion such as Na^+ and n is typically between 9.6 and 12. The fundamental structural unit in X-type zeolites is a truncated octahedron of stoichiometry $M_{24}O_{36}$, where M is Si or Al (Fig. 3-1). A truncated octahedron has six square faces and eight hexagonal faces; the silicon and aluminum atoms occupy the 24 vertexes, and the oxygen atoms lie along each of the 36 edges. For every truncated octahedron in the X-type structure, four of the hexagonal faces are connected to a hexagonal face of an adjacent truncated octahedron by bridging oxygen atoms (Fig. 3-1). This way of connecting the truncated octahedra together produces a solid in which large empty cavities, called supercages, are connected together by apertures composed of 12 oxygen and 12 Si/Al atoms. These 24-atom rings define a pore size for the solid; X-type zeolites have medium-sized pores with diameters of about 7.4 Å.

The reaction you will use to synthesize the X-type zeolite is

$$(24 - n)\,SiO_2 \;+\; n\,NaAlO_2 \;\longrightarrow\; Na_nSi_{24-n}Al_nO_{48}$$

Under the conditions you will use, the product will have a value of n of about 10.7. In this experiment, sodium cations are used to balance the charge, and the formula is often abbreviated NaX. The sodium cations, together with water molecules, fill the supercages. To make a catalyst from NaX, it is necessary to replace Na^+ with H^+ or transition metal cations by means of an ion exchange reaction. In this experiment, you will carry out an exchange reaction with cobalt cations:

$$Na_nSi_{24-n}Al_nO_{48} \;+\; x\,CoCl_2 \;\longrightarrow\; Co_xNa_{n-2x}Si_{24-n}Al_nO_{48} \;+\; 2x\,NaCl$$

Ideally, two Na^+ ions will exchange for every Co^{2+}, as shown in this equation. In reality, a stoichiometric exchange is unlikely because different cobalt species are possible. For example, a Cl^- or OH^- anion might remain coordinated to the

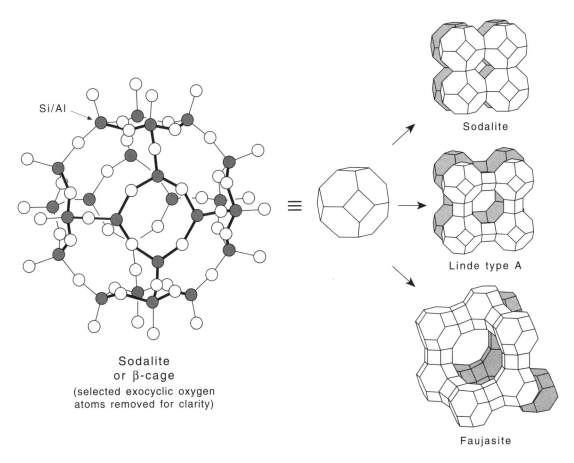

Figure 3-1
Structural relationship between various zeolites derived from the sodalite cage. Zeolite X is
of the faujasite variety. (Figure adapted from Newsam, J. M. in *Solid State Chemistry:
Compounds*, Cheetham A. K.; Day, P., Eds., Oxford University: Oxford, UK, 1992.)

cobalt center, so that, in order to maintain electroneutrality, only one Na^+ ion is
exchanged per cobalt atom introduced.

The synthesis of NaX begins with the preparation of two solutions. The first
is a solution of sodium silicate, Na_2SiO_3, which is prepared from silica gel (a
hydrated form of silicon dioxide) and sodium hydroxide. (For a discussion of the
chemistry of silica gel, see Experiment 23.) The second solution contains sodium
aluminate, $NaAlO_2$, which is prepared from aluminum isopropoxide and sodium
hydroxide. In the presence of excess hydroxide, the aluminate actually exists in
solution in the form of $Al(OH)_4{}^-$ anions. These anions react with sodium silicate
to form the aluminosilicate product. The hydrated sodium cations dictate the size
of the cavities in the aluminosilicate lattice because the lattice grows around
them. The hydrated sodium cation is just the right size to promote growth of
X-type zeolites.

Figure 3-2
A type of autoclave. (Courtesy
of Parr Instrument Co.)

Because NaX is best prepared at temperatures above the boiling point of
water, a special apparatus is needed to contain the high pressures generated.
High-pressure vessels (also called autoclaves or bombs) are frequently used
to carry out reactions in which one or more of the reactants would be a gas
at atmospheric pressure but a liquid under pressure. Autoclaves are also used
to maintain higher concentrations of gases in solution than would be possible at
1-atm pressure. Although in the present experiment a high-pressure autoclave is
not necessary, autoclaves are commonly used to prepare zeolites and many other
materials.

One type of autoclave is shown in Figure 3-2. It is a simple split-ring closure,
general purpose bomb that consists of a cylinder and a head fitted with an inlet–
outlet valve and pressure gauge. The pressure gauge unit also contains a safety
blowout (rupture) disk in case the pressure exceeds the recommended maximum
pressure. Regardless of the specific autoclave used, you should always read the
manufacturer's instructions carefully and understand how to assemble and use it.
Also, you should know its pressure limits.

Often, reaction mixtures are not placed directly into the autoclave but are
put into a liner instead. The liner is simply a large flat-bottom test tube that fits
snugly into the autoclave. Liners can be made of glass, polyethylene, Teflon, or
other materials, and which material is used depends on the reaction temperature
and the nature of the reagents used.

After the the body of the autoclave is charged with the reactants, it is fitted
with a gasket. The head is placed on top of the gasket and is secured to the body
by means of a collar and bolts. Opposite bolts should be tightened alternately (see

the manufacturer's instructions). Then the inlet–outlet valve is closed, and the autoclave is heated to the appropriate temperature (Fig. 3-2). It is a good idea to place a safety screen around the autoclave and to attach a sign warning others to leave the reaction undisturbed. After the reaction period, the autoclave is allowed to cool to room temperature. Any pressure inside the autoclave is released by slowly opening the inlet–outlet valve and allowing the gases to vent into a hood. The head is removed and the solution in the autoclave is collected.

Fortunately, NaX can be prepared even if an autoclave is not available. A modified recipe that works at 90 °C is described below. At this temperature, water remains a liquid and very little pressure is generated. Zeolite NaX is a metastable phase, meaning that other types of zeolites, such as zeolite-A, zeolite-P, or sodalite, may form if the recipe is not followed carefully.

EXPERIMENTAL PROCEDURE

All preparations will be carried out in polypropylene beakers and polypropylene screw-top bottles to prevent contact of the strongly basic solutions with glass. (Polyethylene bottles are not suitable because they will crack and leak.) Mixing is done with Teflon-coated stir bars. *Note*: If this experiment is to be carried out at high altitudes where the boiling point of water is significantly less than 100 °C, then the ambient pressure procedure will not work and instead the synthesis must be carried out in an autoclave.

Part A

Solution 1, Sodium Aluminate
In a 200-mL polypropylene beaker, place 6.9 g (4.9 mmol) of aluminum iso-propoxide, 2.4 g (60 mmol) of sodium hydroxide, and 9 mL of deionized water. Place a watch glass over the beaker to prevent water loss, and heat the mixture to about 70 °C in a water bath until all the solids are dissolved and a clear gel is formed (*do not* heat the mixture above 80 °C!). While the solution is heating, prepare solution 2.

After the solids dissolve, solution 1 contains sodium aluminate, which is generated by the reaction:

$$Al(OC_3H_7)_3 \; + \; NaOH \; + \; 3\,H_2O \; \longrightarrow \; NaAl(OH)_4 \; + \; 3\,HOC_3H_7$$

The isopropanol that forms during the hydrolysis of the aluminum isopropoxide will either evaporate or form droplets at the surface. There is no need to separate the alcohol from the gel.

Solution 2, Sodium Silicate
In a second 200-mL polypropylene beaker, place 3.0 g (45 mmol) of silica gel, 2.4 g (60 mmol) of sodium hydroxide pellets, and 6 mL of deionized water. Swirl the

mixture until the solids are completely dissolved. This solution provides a source of silicate, which is the building block for the framework structure of the zeolite.

Zeolite-X, $Na_nSi_{24-n}Al_nO_{36}$

When Solution 1 has cooled to room temperature, pour Solution 2 into it. Add an additional 27 mL of water and stir the gel with a spatula until the mixture appears homogeneous. Usually, the initial gelatinous mixture will quickly form a suspension of white solids. (You do not need to wait for this to happen; proceed with the next step.) Transfer the mixture to a polypropylene screw cap bottle, screw on the cap, and place the bottle in an oven at about 90 °C.

All students should place their samples in the oven at the same time to minimize the fluctuation in temperature that would result from opening the oven door. The reaction time depends on the time available: The best results are obtained after 3 to 4 h, but reasonably crystalline zeolite-X can be isolated after 2 h.

After 2–4 h in the oven, the sample is cooled to room temperature, and the mixture is transferred to a polypropylene beaker. Add 150 mL of deionized water, and then suction filter the mixture. (*Do not* transfer the high pH mixture directly to a frit or Buchner funnel; contamination may result from dissolution of the frit or filter paper.) Wash the resulting white zeolite crystals with several 100-mL portions of deionized water. Then air-dry or oven-dry the crystals until the next lab period. *Here is a good stopping point for the first day's work.*

Characterize your product by infrared (IR) spectroscopy (this can be done while the cobalt exchange experiment is in progress). Grind the samples briefly with solid potassium bromide (excessive grinding of the samples should be avoided). Press the mixture into a pellet using a pellet press. The most useful region of the IR spectrum is between 400 and 1200 cm^{-1}; the higher frequency bands near 1700 and 3200 cm^{-1} are due to absorbed water. In more highly crystalline samples, the IR bands become narrower.

If possible, determine the X-ray diffraction pattern of your product and compare it with that of an authentic sample of NaX. (See Experiment 2 for a discussion of the X-ray powder diffraction method.) The X-ray diffraction pattern of NaX, which can be indexed to a cubic unit cell, contains significant reflections at 2θ values of 15.4°, 23.3°, and 26.7°. If the sample of zeolite-X is impure, there will be extra peaks between those in the list above. For example, sodalite has peaks at 14.2°, 24.7°, and 43.4°, while cancrinite has peaks at 14.0° and 19.2°. All of these 2θ values are given for Cu $K\alpha$ radiation.

Part B

Cobalt-Exchanged Zeolite-X

In a 250-mL Erlenmeyer flask, dissolve 0.1 g of $CoCl_2 \cdot 6H_2O$ in 100 mL of deionized water. Add 1 g of NaX, heat the slurry to 70 °C, and stir until the pink solution is decolorized. The time it takes for the solution to turn colorless will vary, but at this temperature it should take less than 1 h. Collect the pink precipitate by filtration, wash it with two 50-mL portions of deionized water, and allow it to dry in air. If time permits, heat some of the cobalt-exchanged zeolite in an oven; the powder will eventually turn from pink to blue-violet. The change in color results from dehydration of the cobalt centers. If possible, take UV–visible spectra of both the hydrated and dehydrated samples.

REPORT

Include the following:
1. Percentage yield of NaX.
2. Interpretation of the X-ray powder diffraction pattern of NaX, including the length of the unit cell axis.
3. Infrared spectrum of NaX in a KBr pellet.
4. UV–visible spectrum of the cobalt-exchanged zeolite and assignments of the bands seen (optional).

PROBLEMS

1. From the X-ray pattern of your sample, record the 2θ values of the peaks and calculate the d spacing for each reflection using the Bragg equation (see Experiment 2) and the wavelength of radiation used (Cu $K\alpha = 1.542$ Å; Mo $K\alpha = 0.7107$ Å). Did you synthesize NaX?
2. Describe the roles of sodium hydroxide in this synthesis.
3. The silicon–aluminum framework is negatively charged, and sodium ions are needed to maintain charge neutrality. Replacement of these cations with protons produces an acidic zeolite. Predict what effect the Si/Al ratio will have on this acidity.
4. Explain how the unique structure of zeolites is important in catalytic processes such as the conversion of methanol to gasoline.

INDEPENDENT STUDIES

A. Analyze your sample of NaX by thermogravimetric analysis (TGA) to determine at what temperature the water is driven off and how much weight is lost.
B. Study the absorption of vapors of different molecules into the zeolite lattice to determine the pore size. (Balkus, K. J.; Ly, K. T. *J. Chem. Educ.* **1991**, *68*, 875.)
C. Prepare the "mesoporous" zeolite MCM-41. (Example B in Beck, J. S.; Vartuli, J. C.; Roth, W. J.; Leonowicz, M. E.; Kresge, C. T.; Schmit, K. D.; Chu, C. T.-W.; Olson, D. H.; Sheppard, E. W.; McCullen, S. B.; Higgins, J. B.; Schlender, J. L. *J. Am. Chem. Soc.* **1992**, *114*, 10834. See also, Beck, J. S.; Vartuli, J. C.; Kennedy, G. J.; Kresge, C. T.; Roth, W. J.; Schramm, S. E. *Chem. Mater.* **1994**, *6*, 1816; Edler, K. J.; White, J. W. *J. Chem. Soc., Chem. Commun.* **1995**, 155.)
D. Prepare zeolite-A or zeolite-Y. (Rollmann, L. D.; Valyocsik, E. W. *Inorg. Synth.* **1983**, 22, 61; Blatter, F.; Schumacher, E. *J. Chem. Educ.* **1990**, *67*, 519.)

E. Prepare the zeolite ZSM-5 in a microwave oven. (Arafat, A.; Jansen, J. C.; Ebaid, A. R.; van Bekkum, H. *Zeolites* **1993**, *13*, 162.)

F. Study the growth of Liesegang rings in silica gel. (Sharbaugh, A. H. III; Sharbaugh, A. H. Jr. *J. Chem. Educ.* **1989**, *66*, 589.)

REFERENCES

Zeolites

Balkus, K. J.; Ly, K. T. *J. Chem. Educ.* **1991**, *68*, 875. Synthesis and characterization of NaX.

Barrer, R. M. *Hydrothermal Chemistry of Zeolites*, Academic: New York, 1981. Together with the following book, still the standard texts on zeolites.

Breck, D. W. *Zeolite Molecular Sieves*, Wiley: New York, 1974.

Chen, N. Y.; Degnan, T. F.; Smith, C. M. *Molecular Transport and Reaction in Zeolites*, VCH: New York, 1994. Good textbook on the reactions of molecules inside zeolites.

Gates, B. C. *Catalytic Chemistry*, Wiley: New York, 1992. The use of zeolites as catalysts.

Hamilton, K. E.; Coker, E. N.; Sacco, A. Jr.; Dixon, A. G.; Thompson, R. W. *Zeolites* **1993**, *13*, 645. Effect of silica source on the synthesis of zeolite-X.

Occelli, M. L.; Robson, H. E., Eds, *Zeolite Synthesis*, ACS. Symp. Ser., vol. 398, American Chemical Society: Washington DC, 1989.

Olson, K. H. *Zeolites* **1995**, *15*, 439. X-ray crystal structure of NaX.

Rollmann, L. D. *Adv. Chem. Ser.* **1979**, *173*, 387. Review of zeolites.

Rollmann, L. D.; Valyocsik, E. W. *Inorg. Synth.* **1983**, *22*, 61. Synthesis of zeolite-A, zeolite-Y, offretite, and ZSM-5.

Schwochow, F.; Puppe, L. *Angew. Chem., Int. Ed. Engl.* **1975**, *14*, 620. Overview of the synthesis, structure, and applications of zeolites.

Szostak, A. *Molecular Sieves*, 2nd ed., Blackie: London, 1998. Survey of the synthesis and characterization of aluminosilicate- and phosphate-based zeolitic materials.

Autoclave Methods

Augustine, R. L. *Catalytic Hydrogenation*, Marcel Dekker: New York, 1965, p 3. Description of apparatus and techniques used for high-pressure bomb reactions.

Isaacs, N. S. *Liquid Phase High Pressure Chemistry*, Wiley: New York, 1981.

Part II

MAIN GROUP CHEMISTRY

The Borane–Amine Adduct $BH_3:NH_2C(CH_3)_3$

Note: This experiment requires one 4 hour laboratory period.

Compounds containing boron and hydrogen adopt unusual structures that are not observed for the hydrides of the neighboring element carbon. Representative compounds include B_2H_6, B_4H_{10}, B_5H_9, B_5H_{11}, B_9H_{15}, $B_{10}H_{14}$, $B_{10}H_{10}^{2-}$, $B_{12}H_{12}^{2-}$, and many others. Although these species were initially of interest purely for their curious structures, boron hydrides have become important reagents in organic synthesis. Furthermore, the study of their structures provided the foundation for understanding the bonding in metal clusters and carbocations.

Diborane, B_2H_6, is the simplest of the boranes. It is most conveniently prepared from the reaction of sodium borohydride, $NaBH_4$ (frequently called sodium tetrahydroborate), with BF_3 in ether:

$$3\,NaBH_4 \;+\; 4\,BF_3 \;\longrightarrow\; 3\,NaBF_4 \;+\; 2\,B_2H_6$$

At room temperature, diborane is a gas with the molecular structure shown in Figure 4-1.

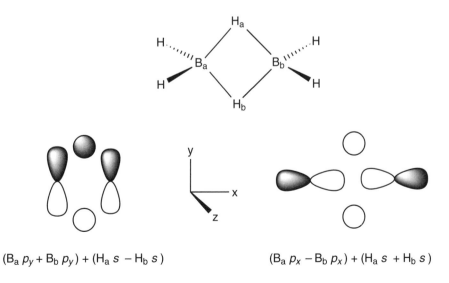

$(B_a\,p_y + B_b\,p_y) + (H_a\,s - H_b\,s)$ $(B_a\,p_x - B_b\,p_x) + (H_a\,s + H_b\,s)$

Figure 4-1
Structure of B_2H_6 and the orbitals involved in B–H–B bonding.

47

The two bridging H atoms lie above and below the plane described by the four terminal H atoms and the two B atoms. Whereas each terminal H atom bonds to only one B atom, each of the bridging H atoms bonds equally to two B atoms. In diborane there are eight B–H bonds, and one might expect that 16 valence electrons are needed to hold the molecule together (normally, a bond consists of an electron pair). Instead, diborane only possesses 12 valence electrons (3 from each boron and 1 from each hydrogen). How are 12 electrons able to form eight B–H bonds? The answer is related to the presence of the bridging hydrogen atoms. The *s* atomic orbitals on these hydrogen atoms overlap with *p* atomic orbitals from both of the adjacent boron atoms to form "multicenter" molecular orbitals as shown in Figure 4-1. The four electrons in these two multicenter molecular orbitals form four B–H bonds, for an average of only *one* electron per bond. The other eight valence electrons in diborane form the normal two-electron bonds to the terminal hydrogen atoms. Such molecular orbitals, consisting of the simultaneous overlap of several atomic orbitals, are common to most of the boron hydrides.

Although B_2H_6 exists largely in the bridged form, it does dissociate to a very small extent to give BH_3.

$$B_2H_6 \rightleftharpoons 2\,BH_3$$

Electron acceptors are sometimes referred to as Lewis acids, and electron donors as Lewis bases. Borane, BH_3, is a Lewis acid because it has six valence electrons and thus can accept two electrons from other molecules. Borane forms numerous complexes (called adducts) with donor molecules such as amines and phosphines. For example, diborane reacts with trimethylamine as shown

$$B_2H_6 + 2\,N(CH_3)_3 \longrightarrow 2\,BH_3{:}N(CH_3)_3$$

The geometry around both the B and N atoms is approximately tetrahedral, and the structure of the adduct is very similar to that of the all-carbon analogue neopentane, $CH_3–C(CH_3)_3$.

Historically, Lewis base adducts of borane were prepared from B_2H_6, a substance that inflames in air and is immediately hydrolyzed by water. In addition to these hazards, B_2H_6 is exceedingly toxic. For these reasons, a more convenient and less dangerous route to the adducts was sought. It was found that the reaction of $NaBH_4$, which is an air stable solid, with an alkylammonium salt produces the borane–amine adduct under mild conditions:

$$NaBH_4 + NHR_3{}^+Cl^- \longrightarrow BH_3{:}NR_3 + NaCl + H_2$$

The particular reaction to be carried out in this experiment involves *tert*-butylammonium chloride; this salt may be simply prepared by bubbling gaseous

HCl into an ether solution of *tert*-butylamine:

$$NH_2C(CH_3)_3 \; + \; HCl \longrightarrow NH_3C(CH_3)_3{}^+Cl^-$$

The borane adduct is formed according to the following equation:

$$NaBH_4 \; + \; NH_3C(CH_3)_3{}^+Cl^- \longrightarrow BH_3 : NH_2C(CH_3)_3 \; + \; NaCl \; + \; H_2$$

The product, $BH_3 : NH_2C(CH_3)_3$, is a white solid (mp 96 °C) that is stable toward air and water at room temperature. You will measure its IR and ^1H nuclear magnetic resonance (NMR) spectra. The ^1H NMR spectrum of $BH_3 : NH_2C(CH_3)_3$ consists of one sharp peak corresponding to the $-CH_3$ protons at approximately $\delta - 1.2$ relative to tetramethylsilane. The corresponding signals for the protons on the N and B atoms are broadened so greatly by the nuclear quadrupole moments of ^{14}N and ^{11}B that they may not be visible in the spectrum.

Borane adducts undergo many reactions. For example, the replacement of H by Cl in the BH_3 portion of the molecule can be accomplished by reaction with gaseous HCl

$$BH_3 : NH_2C(CH_3)_3 \; + \; HCl \longrightarrow BH_2Cl : NH_2C(CH_3)_3 \; + \; H_2$$

This reaction again illustrates the hydridic, H^-, nature of the H atoms attached to the electropositive boron atom. The hydridic H atoms readily combine with protonic, H^+, hydrogen atoms to produce H_2. The analogous reaction with HF replaces all three hydrogen atoms on the boron

$$BH_3 : NH_2C(CH_3)_3 \; + \; 3\,HF \longrightarrow BF_3 : NH_2C(CH_3)_3 \; + \; 3\,H_2$$

The same adduct can also be prepared directly by treating BF_3 with $NH_2C(CH_3)_3$.

Trialkylamine-boranes react with alkenes to form products in which the B and H have added across the double bond

$$BH_3 : N(CH_3)_3 \; + \; 3\,H_2C{=}CHR \longrightarrow B(CH_2CH_2R)_3 \; + \; N(CH_3)_3$$

The trialkylborane $B(CH_2CH_2R)_3$ is a sufficiently weak Lewis acid that it may be liberated from the trimethylamine without difficulty. The ability of a borane to add across the double bond of an alkene in an anti-Markownikov fashion is the basis for the widespread use of boranes in organic chemistry.

Borane-amines lose H_2 at high temperatures and generate a variety of products, depending on the particular reactant and the conditions. When the borane-methylamine adduct, $BH_3 : NH_2(CH_3)$, is heated to 100 °C, it yields $B_3N_3H_3(CH_3)_3$, which is the B–N analogue of 1,3,5-trimethylcyclohexane. Further loss of H_2 occurs at roughly 200 °C to produce the unsaturated cyclic ring compound $B_3N_3H_3(CH_3)_3$ called 1,3,5-trimethylborazole:

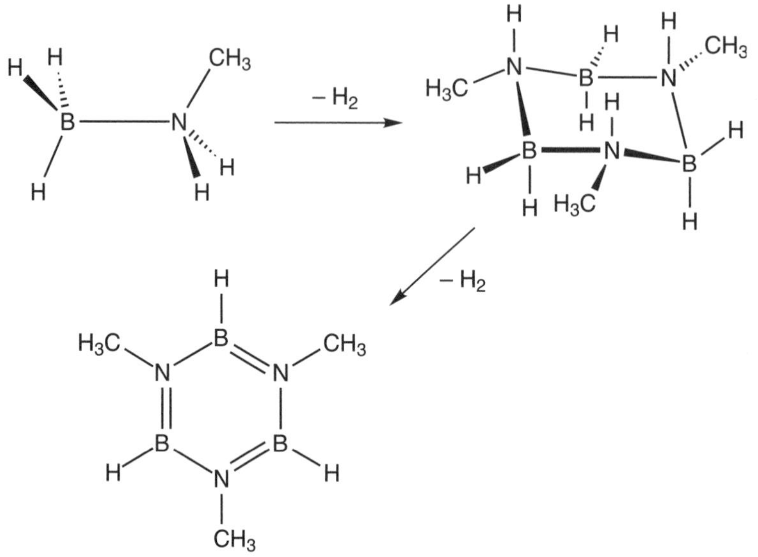

The carbon analogue of 1,3,5-trimethylborazole is mesitylene, $1,3,5-C_6H_3(CH_3)_3$. The parent compound borazole, $B_3N_3H_6$, structurally resembles the isoelectronic compound benzene. For this reason, borazole is sometimes called "inorganic benzene." Like benzene, it has a planar hexagonal structure, and the relatively short B–N bonds indicate the presence of B–N π bonding. Despite the physical similarities, benzene and borazole have quite different chemical reactivities. The π system of benzene is relatively inert to addition reactions. In contrast, borazole adds hydrogen halides, HX, to give the saturated cyclic aminoborane $B_3N_3H_9X_3$. The susceptibility of borazole to attack is related to the polar nature of the B–N bond.

EXPERIMENTAL PROCEDURE

Note: In contrast to many hydride compounds, such as NaH, CaH_2, or $LiAlH_4$, which react explosively with water, $NaBH_4$ is stable in neutral or alkaline aqueous solutions. It rapidly hydrolyzes in acidic solution, however. Fresh $NaBH_4$ should be used in this experiment—old samples of $NaBH_4$ will give inconsistent or poor yields.

tert-Butylammonium Chloride, $NH_3C(CH_3)_3{}^+Cl^-$

In a hood, dissolve 2.5 mL (1.7 g, 23 mmol) of 2-amino-2-methylpropane [*tert*-butylamine, $NH_2C(CH_3)_3$] in 20 mL of anhydrous diethyl ether. Cautiously bubble gaseous HCl from a compressed gas cylinder into the solution until precipitation of $NH_3C(CH_3)_3{}^+Cl^-$ is complete. Suction filter the product on a medium frit, wash the solid with a few milliliters of ether, and dry the solid in a vacuum. Determine the yield. Although many alkylammonium chloride salts are very hygroscopic, NH$_3$C(CH$_3$)$_3$$^+Cl^-$ is not; it need not be stored in a desiccator except when the humidity is high.

tert-Butylamine-Borane, BH$_3$:NH$_2$C(CH$_3$)$_3$

Assemble the apparatus shown in Figure 4-2 and lubricate the stirring shaft bearing with glycerin. The drying tube is necessary only if the atmospheric

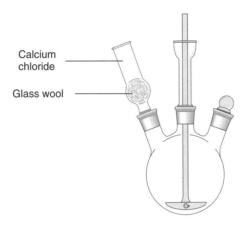

Calcium chloride

Glass wool

Figure 4-2
Apparatus for preparation of BH$_3$:NH$_2$C(CH$_3$)$_3$.

humidity is very high. Add 1.3 g (11.8 mmol) of NH$_3$C(CH$_3$)$_3$$^+Cl^-$ and 15 mL of tetrahydrofuran (THF) to the 250-mL three-neck flask. (The THF may be used as obtained commercially unless it contains large amounts of water. Then it should be dried over NaOH and distilled.) To the stirred suspension, add 0.20 g (5.3 mmol) of powdered NaBH$_4$. At this point, H$_2$ gas will be evolved. Add an additional 10–15 mL of THF and continue stirring the solution for about 2 h at room temperature. Filter the solution using a suction filtration apparatus (see Fig. 13-1). After the filtration, disconnect the rubber vacuum tubing from the filter flask *before* the water flow is turned off; this action will prevent water from backing up into the trap. Discard the solid (which contains NaCl and unreacted excess NH$_3$C(CH$_3$)$_3$$^+Cl^-$) and keep the solution. Using a rotary evaporator, evaporate the THF solution to dryness. The BH$_3$:NH$_2$C(CH$_3$)$_3$ product that remains in the flask is usually of high purity; determine its melting point to confirm this. If the melting range is less than 4 °C, skip the following recrystallization step. If the melting range is greater than 4 °C, the compound should be recrystallized by dissolving it in a minimum amount (1–2 mL) of toluene and adding 20 mL of hexane until precipitation is complete. Collect the BH$_3$:NH$_2$C(CH$_3$)$_3$ by suction filtration and dry the solid product in air. Redetermine its melting point.

Finally, calculate the yield of product. Measure the IR spectrum of the product in CHCl$_3$ or CDCl$_3$ solution.

REPORT

Include the following:
1. Percentage yields of $NH_3C(CH_3)_3{}^+Cl^-$ and $BH_3:NH_2C(CH_3)_3$.
2. Melting point of $BH_3:NH_2C(CH_3)_3$.
3. Infrared spectrum of $BH_3:NH_2C(CH_3)_3$ with assignments to vibrational modes in the molecule. Compare the B–H, C–H, and N–H stretching frequencies and account for their differences.

PROBLEMS

1. Propose a mechanism for the reaction of $NaBH_4$ with $NH_3C(CH_3)_3{}^+Cl^-$.
2. Suggest a method of establishing the presence of boron in your product, $BH_3:NH_2C(CH_3)_3$.
3. Earlier it was noted that B_2H_6 inflames in air and rapidly hydrolyzes in water. Write balanced equations for these reactions.
4. Write a balanced equation for the hydrolysis of $NaBH_4$ in acidic solution.
5. Draw structures of the following: $NaBF_4$, $NaBH_4$, $B(CH_3)_3$, and BF_3.
6. Tetrahydrofuran forms an adduct with BH_3, $BH_3:THF$. Draw the structure of this compound. Is there any evidence from this experiment that would suggest that THF coordinates more or less strongly to BH_3 than does $NH_2C(CH_3)_3$?
7. Account for the fact that $LiBH_4$ is more soluble in THF than is $NaBH_4$. Would you expect $LiBH_4$ or $NaBH_4$ to give better yields in the present experiment? Why?
8. If you wished to carry out a reaction of the type

$$BH_3:NH_2C(CH_3)_3 + amine \longrightarrow BH_3:amine + NH_2C(CH_3)_3,$$

what amine would you choose and what reaction conditions would you use to drive the reaction to completion?

INDEPENDENT STUDIES

A. Prepare and characterize other amine-borane adducts, such as $BH_3:NH(CH_3)_2$ and $BH_3:N(CH_3)_3$. (Nöth, H.; Beyer, H. *Chem. Ber.* **1960**, *93*, 928. Nainan, K. C.; Ryschkewitsch, G. E. *Inorg. Synth.* **1974**, *15*, 122.)
B. Using your $BH_3:NH_2C(CH_3)_3$, prepare and characterize $BH_2X:NH_2C(CH_3)_3$ (where X = F, Cl, Br, or I). (Nöth, H.; Beyer, H. *Chem. Ber.* **1960**, *93*, 2251. Ryschkewitsch, G. E.; Wiggins, J. W. *Inorg. Synth.* **1970**, *12*, 116.)
C. Prepare and characterize $BH_3:py$ (where py = pyridine) and $[BH_2(py)_2{}^+]I^-$. (Ryschkewitsch, G. E. *J. Am. Chem. Soc.* **1967**, *89*, 3145. Nainan, K. C.; Ryschkewitsch, G. E. *Inorg. Chem.* **1968**, *7*, 1316.)
D. Determine the mass spectrum of $BH_3:NH_2C(CH_3)_3$ and make assignments to all ion fragments.
E. Prepare the interesting tridentate ligand, hydrotris(1-pyrazolyl)borate, $HB(C_3H_3N_2)_3{}^-$. (Trofimenko, S. *Inorg. Synth.* **1970**, *12*, 102.)

REFERENCES

$BH_3:NH_2R$

Nöth, H.; Beyer, H. *Chem. Ber.* **1960**, *93*, 928. Preparation and characterization of borane adducts.

Paetzold, P. *Adv. Inorg. Chem.* **1987**, *31*, 12. Review of imido boranes.

Paine, R. T.; Narula, C. K. *Chem. Rev.* **1990**, *90*, 73. Synthesis and thermal decomposition of B–N compounds.

Taylor, R. C. in *Boron–Nitrogen Chemistry*, Advances in Chemistry Series, No. 42, American Chemical Society: Washington DC, 1964, p 59. Infrared spectra of borane adducts.

Boron Hydride Chemistry

Brown, H. C. *Organic Syntheses via Boranes*, Wiley: New York, 1980.

Grimes, R. N., Ed. *Metal Interactions with Boron Clusters*, Plenum: New York, 1982.

Housecroft, C. E. *Boranes and Carboranes*, 2nd ed., Ellis Horwood: New York, 1994.

Muetterties, E. L.; Knoth, W. H. *Polyhedral Boranes*, Marcel Dekker: New York, 1968. Structure, bonding, and reactions of the higher boranes.

Sneddon, L. G.; Wideman, T. *Inorg. Chem.* **1995**, *34*, 1002. Large-scale synthesis of borazole.

Stock, A.; Pohland, E. *Chem. Ber.* **1926**, *59*B, 2215. First synthesis of borazole.

Trofimenko, S. *Chem. Rev.* **1993**, *93*, 943. Review of poly(pyrazolyl)borates.

Williams, R. E. *Adv. Organomet. Chem.* **1994**, *36*, 1. A review of carborane structures and electron counting rules.

Woollins, J. D. *Non-Metal Rings, Cages, and Clusters*, Wiley: New York, 1988.

General References to Main Group Chemistry

King, R. B. *Inorganic Chemistry of Main Group Elements*, VCH: New York, 1995.

Massey, A. G. *Main Group Chemistry*, Ellis Horwood: Chichester, 1990.

Steudel, R. *Chemistry of the Non-Metals, with an Introduction to Atomic Structure and Chemical Bonding*, W. de Gruyter: New York, 1977.

Buckminsterfullerene (C_{60}) and Its Electrochemistry

Note: This experiment requires 4 hours spread over two laboratory periods.

The two most common and stable forms of carbon are diamond and graphite. In diamond the carbon atoms are tetrahedral (sp^3-hybridized) and form a three-dimensional network, whereas in graphite the carbon atoms are trigonal planar (sp^2-hybridized) and form a two-dimensional network consisting of layers of six-membered rings. Several other forms of carbon have been discovered more recently, the best known being the highly symmetrical species C_{60}. This molecule is also known as Buckminsterfullerene because its structure is reminiscent of the domed buildings designed by the architect Buckminster Fuller. The entire class of inorganic carbon cages, called *fullerenes*, includes C_{60}, C_{70}, C_{76}, C_{84}, and some still larger species. Because the fullerenes are molecular species rather than infinite polymers, they are soluble in organic solvents.

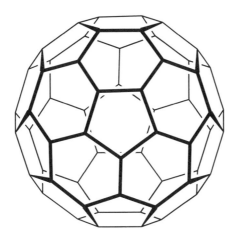

Like graphite, C_{60} is constructed from three-coordinate carbon atoms, but unlike graphite there are both five- and six-membered rings. The structure of C_{60} is a truncated icosahedron, a polyhedron with 12 five-membered rings and 20 six-membered rings. Buckminsterfullerene has very high symmetry: It has 6 fivefold

55

rotation axes as well as 10 threefold rotation axes. All 60 carbon atoms are equivalent because they can be interchanged by the symmetry operations. The very high symmetry of C_{60} is indicated by the presence of a single resonance in its ^{13}C nuclear magnetic resonance (NMR) spectrum. The position of this resonance, at $\delta\,142.3$, is in the range expected for aromatic compounds such as benzene ($\delta\,128.5$), but is shifted owing to the pyramidalization forced upon the carbon atoms by the cage structure.

In addition to its beautiful structure, C_{60} has many interesting chemical properties. For example, the cage can be reduced to give a series of C_{60}^{n-} anions:

$$C_{60} \underset{-e^-}{\overset{+e^-}{\rightleftharpoons}} C_{60}^- \underset{-e^-}{\overset{+e^-}{\rightleftharpoons}} C_{60}^{2-} \underset{-e^-}{\overset{+e^-}{\rightleftharpoons}} C_{60}^{3-} \underset{-e^-}{\overset{+e^-}{\rightleftharpoons}} C_{60}^{4-}$$

The cage can also be alkylated with organolithium reagents and it can be oxygenated, hydrogenated, and halogenated to give derivatives such as $C_{60}O$, $C_{60}H_2$, and $C_{60}F_{28}$. Consistent with its ability to accept electrons, C_{60} acts like an electron-poor polyalkene. Electron-poor alkenes are well known to coordinate to low valent transition metal complexes and C_{60} is no exception. Representative metal complexes include $C_{60}IrCl(CO)[P(C_6H_5)_3]_2$ and $C_{60}\{Pt[P(C_2H_5)_3]_2\}_6$. As shown by the latter complex, several metals can coordinate to the cage simultaneously, a limiting factor being steric interactions between the appended metal centers. Finally, the cage can also react with dienes to give Diels–Alder adducts. Virtually all of the reactions of C_{60} occur at the fusion of pairs of six-membered rings; partial structures of some of the adducts are shown below:

The synthesis of C_{60} begins with the evaporation of carbon. With the help of a welder's apparatus, a large electrical current (called an arc discharge) is established across a gap separating two graphite rods. The arc discharge heats the graphite, causing some of the carbon to evaporate. When the arc discharge is carried out through helium gas at low pressure, soot is generated that contains a few percent of C_{60} and trace amounts of the other fullerenes. The fullerenes can be separated from the insoluble reaction products by washing the soot with suitable solvents. The solubility of the fullerenes is not very high, however, and the extraction process can be slow. One can extract the C_{60} by repeatedly washing the soot with the chosen solvent, but this is inefficient and wasteful. The use of a Soxhlet extractor solves this problem.

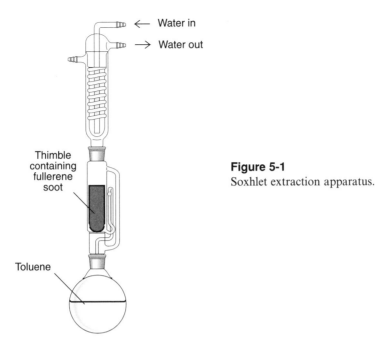

← Water in

→ Water out

Thimble
containing
fullerene
soot

Toluene

Figure 5-1
Soxhlet extraction apparatus.

A Soxhlet extractor consists of a glass reservoir into which is placed a paper thimble containing the material to be extracted. Like filter paper, the thimble is permeable to the solvent but retains solids. The extractor containing the thimble is attached both to a flask charged with the extracting solvent and to an efficient condenser (Fig. 5-1). When the solvent is brought to a vigorous boil, the solvent vapors travel up the large side arm of the extractor. It is often necessary to wrap the large side arm on the Soxhlet extractor with glass wool or aluminum foil, especially when using less volatile solvents. This insulation prevents the boiling solvent from condensing prematurely in the side arm. When the vapor contacts the condenser, the solvent recondenses and drips onto the sample in the thimble, where it extracts the solid. It is important to use a sufficient amount of the extracting solvent so that the solvent flask does not go dry when the Soxhlet reservoir becomes filled. Once the extractor becomes filled to the level of its siphon side arm, the solution drains back into the solvent reservoir and the cycle begins anew. Because the extracted material is typically nonvolatile, it remains in the solvent flask. Extraction is allowed to continue so long as the material in the thimble is leached out. Even poorly soluble compounds can be purified in this way.

Once the extraction is complete, the next step is to separate the C$_{60}$ from the other fullerenes, which are usually present in small amounts. This separation can be achieved by column chromatography—that is, by pouring the extract through a column containing a mixture of silica gel and charcoal. Silica gel is a common stationary phase for chromatography, but it interacts only weakly with the fullerenes. Charcoal interacts strongly with fullerenes (an extension of the concept that like dissolves like), but it does not pack as well as silica gel in the column. A mixture of the two seems to be an effective compromise. In this experiment, the fullerenes travel down the column as a broad violet-red band. The initial fractions of the band are violet and contain pure C_{60}, whereas the later fractions are red and contain a mixture of C_{60}, C_{70}, and trace amounts of higher fullerenes. More detailed information on chromatography is provided in Experiment 23.

Cyclic Voltammetry

The redox properties of a compound can be readily analyzed by means of a technique known as cyclic voltammetry (CV). In voltammetric measurements, the compound to be studied is first dissolved in a suitable solvent. This solution is subjected to a variable electric potential (E) and the current that flows through the solution (i) induced by this electric potential is monitored. The electric potential is generated by an external power supply connected to two electrodes, a working electrode and a counterelectrode, dipped into the solution (Fig. 5-2).

In a typical cyclic voltammogram, there is a range of potentials at which the electrode is unable to donate electrons to or accept electrons from any species in solution. Thus, within this range of potentials, no current is passed. Any voltage within this range is called a "rest potential". Within the rest potential range, the solution will have the *same* chemical composition at the electrode surface as it does in the solution as a whole. When the potential, measured in volts (V), is moved outside the rest potential range and approaches the redox potential ($E_{1/2}$) of the complex, molecules near the surface of the electrode begin to be oxidized or reduced. A burst of current then flows to or from the working electrode. Because the solution is not stirred, the current soon decreases as the molecules

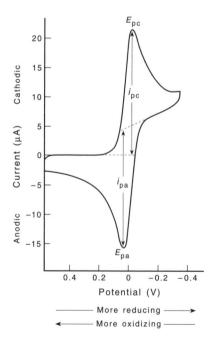

Figure 5-2

Schematic of a cyclic voltammetry experiment.

near the working electrode become oxidized or reduced. The current does not decrease to zero, however, because diffusion processes slowly bring fresh molecules to the electrode. If one plots the current as a function of the electrical potential, the plot consists of a horizontal line that rises to a peak and then falls to another horizontal line at a higher level.

Because CV involves the passage of current, the solution needs to be electrically conductive. In most cases, this is accomplished by adding to the solution an unreactive salt called a supporting electrolyte. In nonaqueous solvents, quaternary ammonium salts such as $N(C_4H_9)_4{}^+PF_6{}^-$ are customarily used as supporting electrolytes.

In *cyclic* voltammetry, the i versus E curve is recorded twice, once in the oxidizing direction and once in the reducing direction. One can equally well start with a reductive scan followed by an oxidative one. All the molecules near the electrode surface that have been oxidized (or reduced) in the first sweep are given an opportunity to be rereduced (or reoxidized) in the reverse scan. If the currents for the peaks, i_p, in the two scans match, then the redox process is reversible and can probably be accomplished chemically using oxidants or reductants that are chosen on the basis of their redox potential. If the currents do not match, the process is irreversible, which usually means that the electrogenerated product is chemically unstable. For example, the reduction of aqueous $Co(NH_3)_6{}^{3+}$ is irreversible because $Co(NH_3)_6{}^{2+}$ readily loses its NH_3 ligands in aqueous solution. Thus, soon after its formation, the $Co(NH_3)_6{}^{2+}$ converts to $Co(OH_2)_6{}^{2+}$, which does not reoxidize at the same potential as $Co(NH_3)_6{}^{2+}$. By varying the scan rate and determining its effect on the relative magnitude of the peak in the oxidizing direction (called the anodic current, i_{pa}) and the peak in the reducing direction (called the cathodic current i_{pc}), one can deduce rate constants for a decomposition process. The measurement of peak currents requires that one take account of the background current; a method for doing this is shown in Figure 5-3.

The potentials at which the maximum (peak) currents are observed are called E_p values. The potentials for the anodic (oxidizing) and cathodic (reducing) sweep directions are called E_{pa} and E_{pc}, respectively. The potential $E_{1/2}$ for an electrochemical process is equal to $(E_{pa} + E_{pc})/2$. The $E_{1/2}$ determined in this manner corresponds to the redox potential of the complex provided that the diffusion properties, that is, the mobilities, of the reduced and oxidized species are similar (which is usually the case). In the ideal case, the separation of the peak potentials, $\Delta E_p = |E_{pa} - E_{pc}|$, is 0.059 V for a one-electron redox process. For nonaqueous solutions, this ideal is rarely achieved, usually because of the high electrical resistance of the solution. It is common to assess the reversibility of a redox process by comparing ΔE_p values of the unknown and a reference couple known to be reversible. A commonly employed reference couple is $Fe(C_5H_5)_2{}^{+/0}$.

The cyclic voltammogram of $Fe(C_5H_5)_2$ in acetonitrile shows that a redox process occurs at $E_{1/2} = 0.697$ V versus the standard hydrogen electrode (SHE; this reference electrode will be discussed below). At potentials between about -1.5 V and about 0.4 V, no current flows either from or to the electrode. Within this rest potential range, the solution near the electrode surface contains the same species, $Fe(C_5H_5)_2$, as does the bulk solution. In contrast, as the potential approaches the $E_{1/2}$ value of 0.697 V, the electrode begins accepting electrons

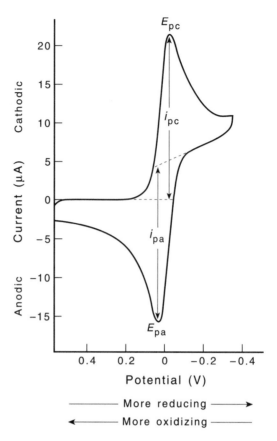

Figure 5-3
Cyclic voltammogram for a reversible one-electron
redox couple with $E_{1/2} = 0$ V.

from the solution. Thus, as one sweeps the potential to higher voltages (called the anodic direction), $Fe(C_5H_5)_2$ is being oxidized. Careful measurements of the amount of current passed show that exactly one electron is lost and the cation $Fe(C_5H_5)_2^+$ is formed

$$Fe(C_5H_5)_2{}^0 \rightleftharpoons Fe(C_5H_5)_2{}^+ + 1\ e^-$$

When the potential is swept back to lower voltages (called the cathodic direction), the $Fe(C_5H_5)_2^+$ cations generated in the first sweep are reduced back to $Fe(C_5H_5)_2$. The peak currents in the initial and final potential sweeps of the redox process are almost identical, so this redox process is reversible. Because both $Fe(C_5H_5)_2{}^0$ and $Fe(C_5H_5)_2{}^+$ are stable for the duration of the sweep, it should be possible to oxidize $Fe(C_5H_5)_2{}^0$ by adding a chemical oxidant. The $E_{1/2}$ for this process suggests that $Fe(C_5H_5)_2{}^0$ can be oxidized by chlorine ($E_{1/2} = 1.36$ V) but not by iodine ($E_{1/2} = 0.535$ V).

Most cyclic voltammetry experiments employ three electrodes: a working electrode, where the redox-active species in solution is oxidized or reduced; a reference electrode, to which the voltage of the working electrode is compared; and a counterelectrode, which balances the current passed into (or out of) the solution at the working electrode. This three electrode design, which minimizes the passage of current through the reference electrode, is advantageous because currents impair the ability of an electrode to function as a reference.

It is important to take note of the reference electrode used in a given experiment and to make corrections when comparing redox potentials. The SHE consists of a platinum wire immersed in an acid (H^+) solution under an H_2 atmosphere. The concentration of acid and the pressure of H_2 are strictly defined: they are close to 1 M and 1 atm, respectively. Most redox potentials are measured relative to the SHE potential. A list of common reference electrodes is given in Table 5-1. From the table you can compute, for example, that the $E_{1/2}$ for $Fe(C_5H_5)_2^{+/0}$ in acetonitrile is 0.492 V referenced to Ag/AgCl.

Table 5-1
Reference Electrodes[a]

Electrode	Potential vs. SHE (V)	Name of Electrode
H_2/H^+	0.000	Standard hydrogen electrode (SHE)
Ag/AgCl	0.205	Silver/silver chloride
Hg/Hg_2Cl_2	0.250	Mercury/mercurous chloride or calomel electrode

[a] Potentials given are for 25 °C. For the silver/silver chloride and calomel electrodes, the potential given is for a 3.5 M KCl electrolyte solution.

EXPERIMENTAL PROCEDURE

Extraction of Fullerenes from Soot
Safety note: toluene and *o*-dichlorobenzene are suspected carcinogens.

Load 2.0 g of C_{60}-bearing soot into a Soxhlet thimble (3 × 15 cm) and cover the powder with a plug of glass wool. Place the thimble in a Soxhlet extraction apparatus fitted onto a 200- or 250-mL round-bottom flask that contains 125 mL of toluene and a stir bar or boiling chip. Place a reflux condenser on top of the extraction apparatus. Grease all joints with silicone stopcock grease and clamp the ground glass joints together securely. Heat the toluene until the solution refluxes and the toluene vapor travels up the Soxhlet side arm and reaches the condenser. Continue the extraction for 3 h or until the extracts are almost colorless. Remove the flask from the extractor and evaporate the brown solution to dryness on a rotary evaporator. Weigh the crude C_{60} and record its infrared (IR) spectrum.

The following operations can be carried out while the Soxhlet extraction is proceeding. To a ring stand, clamp a chromatography column measuring approximately 10 cm in length and 4 cm in diameter. Place a small plug of glass wool at the bottom of the column. In succession, add enough sand to form a 1-cm layer, and then enough silica gel to form a 1-cm layer on top of the sand. In a 100-mL beaker, prepare a slurry from the following: 2.5 g of activated charcoal, 5 g of silica gel, 20 mL of toluene, and 20 mL of o-dichlorobenzene. Close the stopcock on the column and pour the charcoal–silica gel slurry into the column. Open the stopcock and allow the solvent to drain until the liquid level is near the top of the column. Cover the column with aluminum foil until the next laboratory period. *Here is a good stopping point for the first day's work.*

Chromatography of the Crude C_{60}

Suspend the crude C_{60} in a mixture of 5 mL of toluene and 5 mL of o-dichlorobenzene (not all of the crude C_{60} needs to dissolve) and pour the suspension onto the silica gel–charcoal column. Open the valve on the column and allow this solution to drain until there is no more liquid above the top of the silica gel–charcoal bed. Add more of the 1:1 (by volume) mixture of toluene and o-dichlorobenzene, recycling the solvent until the violet band starts to elute, and then collect the violet fraction. After you have collected about 25 mL of the violet C_{60} fraction, or when the eluent color becomes noticably red due to the presence of some C_{70}, stop the collection of this fraction. (If you have time, you may collect a second fraction rich in C_{70} by eluting the column with pure o-dichlorobenzene. Ask your instructor.) To isolate solid C_{60}, evaporate the violet fraction first with a rotary evaporator to remove the toluene and later with a vacuum pump (use a trap!) to remove the o-dichlorobenzene. Weigh your purified product and record its IR spectrum.

Cyclic Voltammetry of C_{60}

Consult your instructor on the operation of the cyclic voltammetry apparatus. To prepare a solution that is 10^{-3} M in C_{60} and 10^{-1} M in the supporting electrolyte, dissolve 0.387 g (1 mmol) of $N(C_4H_9)_4PF_6$ in 10 mL of o-dichlorobenzene. Place some of the solution in the cyclic voltammetry cell and purge the stirred solution with a N_2 gas stream. Add 6 mg (8 mmol) of C_{60} to the solution. After all of the C_{60} has dissolved, turn off the stirrer and record the voltammogram between 0.0 and -2 V versus Ag/AgCl. If you are unable to record the cyclic voltammogram for your compound, analyze the voltammogram given in the paper by Suzuki, which is listed among the references at the end of this experiment.

REPORT

Include the following:
1. Weights of the crude and purified samples of C_{60}.
2. Infrared spectrum of C_{60}.
3. Cyclic voltammogram of C_{60}.

PROBLEMS

1. How many ^{13}C NMR resonances would you expect for $C_{60}O$ and in what relative intensity?
2. Would you expect C_{60} to form a stronger complex with $IrCl(CO)[P(CH_3)_3]_2$ or $IrCl(CO)[P(C_6H_5)_3]_2$? Which of the two alkenes, C_2H_4 or C_2FH_3, more closely resembles C_{60} in terms of its coordination properties?
3. The first reduction of C_{60} occurs at -0.450 V versus SHE and the second reduction (to C_{60}^{2-}) occurs at -0.800 V. What reducing agents would be suitable for the reduction of C_{60} to C_{60}^- without formation of C_{60}^{2-}? (See Subramanian, R.; Boulas, P.; Vijayashree, M. N.; D'Souza, R. D.; Jones, M. T.; Kadish, K. M. *J. Chem. Soc., Chem. Commun.* **1994**, 1847.)
4. Suppose that from your chromatography experiment you obtain what you believe to be a new fullerene. What methods would you employ to identify your new material?

INDEPENDENT STUDIES

A. Isolate C_{70} by continuing the chromatography step using pure *o*-dichlorobenzene. The C_{70} is reddish in color. (Scrivens, W. A.; Bedworth, P. V.; Tour, J. M. *J. Am. Chem. Soc.* **1992**, *114*, 7917.)
B. Examine the reaction of C_{60} with $IrCl(CO)[P(C_6H_5)_3]_2$. (Balch, A. L.; Lee, J. W.; Noll, B. C. *Inorg. Chem.* **1994**, *33*, 5238. See Experiment 19 for the preparation of Vaska's complex.)
C. Examine the reaction of C_{60} with pentamethylcyclopentadiene. (Meidine, M. F.; Avent, A. G.; Darwish, A. D.; Kroto, H. W.; Ohashi, O.; Taylor, R.; Walton, D. R. M. *J. Chem. Soc., Perkin Trans.* **1994**, 1189.)
D. Investigate the use of mineral oil to extract C_{60} from soot. (West, S. P.; Poon, T.; Anderson, J. L.; West, M. A.; Foote, C. S. *J. Chem. Educ.* **1997**, *74*, 311.)
E. Use a spectroelectrochemical cell to measure the UV–vis spectra of C_{60} and its reduction products C_{60}^{n-}. (Dubois, D.; Kadish, K. M.; Flanagan, S.; Haufler, R. E.; Chibante, L. P. F.; Wilson, L. J. *J. Am. Chem. Soc.* **1991**, *113*, 4364.)

REFERENCES

C_{60} and Other Fullerenes

Craig, N. C.; Gee, G. C.; Johnson, A. R. *J. Chem. Educ.* **1992**, *69*, 664. Preparation of C_{60} and C_{70}.

Curl, R. F.; Kroto, H.; Smalley, R. E. *Angew. Chem., Int. Ed. Engl.* **1997**, *36*, 1567, 1579, 1595. Personal accounts of the discovery of C_{60}.

Haddon, R. C. *Science* **1993**, *261*, 1545. Bonding in the fullerenes.

Hare, J. P.; Dennis, T. J.; Kroto, H. W.; Taylor, R.; Allaf, A. W.; Balm, S.; Walton, D. R. W. *J. Chem. Soc., Chem. Commun.* **1991**, 412. The IR spectra of C_{60} and C_{70}.

Iacoe, A. W.; Potter, W. T.; Teeters, D. *J. Chem. Educ.* **1992**, *69*, 663. Preparation of C_{60}.

Krätschmer, W.; Lamb, L. D.; Fostiropoulos, K.; Huffman, D. R. *Nature (London)* **1990**, *347*, 354. Isolation of macroscopic amounts of C_{60}.

Kroto, H. W.; Heath, J. R.; O'Brien, S. C.; Curl, R. F.; Smalley, R. E. *Nature (London)* **1985**, *318*, 162. The original mass spectrometric evidence for large carbon cages and the proposal that C_{60} adopts a soccer ball-like structure with icosahedral symmetry.

Derivatives of Fullerenes

Balch, A. L.; Lee, J. W.; Noll, B. C. *Inorg. Chem.* **1994**, *33*, 5238. Binding of Vaska's complex to C_{60}.

Diederich, F.; Thilgen, C. *Science* **1996**, *271*, 317. Organic reactions of fullerenes.

Fagan, P. J.; Calabrese, J. C. *Science* **1991**, *252*, 1160. Synthesis of $Pt(PR_3)_2$ complexes of C_{60}.

Hawkins, J. M. *Acc. Chem. Res.* **1992**, *25*, 150. Addition of OsO_4 to C_{60}.

Hebard, A. F.; Rosseinsky, M. J.; Haddon, R. C.; Murphy, D. W.; Glarum, S. H.; Palstra, T. T. M.; Ramirez, A. P.; Kortan, A. R. *Nature (London)* **1991**, *350*, 600. Discovery of superconductivity in K_3C_{60}.

Henderson, C. C.; Cahill, P. A. *Science* **1993**, *259*, 1885. Synthesis of $C_{60}H_2$.

Hirsh, A.; Lamparth, I.; Karfunkel, H. R. *Angew. Chem., Int. Ed. Engl.* **1994**, *33*, 437. Stereochemistry of multiple additions to C_{60}.

Hsu, H.-F.; Shapley, J. R. *J. Am. Chem. Soc.* **1996**, *118*, 9192. Synthesis of $Ru_3(CO)_9(C_{60})$, a compound in which one face of C_{60} is bound to a metal cluster.

Lappas, A.; Prassides, K.; Vavekis, K.; Arcon, D.; Blinc, R.; Cevc, P.; Amato, A.; Feyerherm, R.; Gygax, F. N.; Schenk, A. *Science* **1995**, *267*, 1799. Superconductivity in charge-transfer salts of C_{60}.

Chromatography

Scrivens, W. A.; Bedworth, P. V.; Tour, J. M. *J. Am. Chem. Soc.* **1992**, *114*, 7917. Chromatographic separation of fullerenes.

Zumwalt, M. C.; Denton, M. B. *J. Chem. Educ.* **1995**, *72*, 939. High-performance liquid chromatography (HPLC) analysis of fullerenes. Further references on chromatography can be found at the end of Experiment 23.

Cyclic Voltammetry

Bard, A. J.; Faulkner, L. R. *Electrochemical Methods: Fundamentals and Applications*, Wiley: New York, 1980. A comprehensive text.

Heineman, W. T.; Kissinger, P. T. *Am. Labor.* **1982**, 29. Introduction to CV.

Heinze, J. *Angew. Chem., Int. Ed. Engl.* **1984**, *23*, 831. A review of CV with a discussion of systems in which chemical and electrochemical processes are coupled.

Kissinger, P. T.; Heineman, W. R., Eds. *Laboratory Techniques in Electroanalytical Chemistry*, Marcel Dekker: New York, 1984. Overview of methods used in electrochemistry.

Sawyer, D. T.; Sobkowiak, A.; Roberts, Jr., J. L. *Experimental Electrochemistry for Chemists*, 2nd ed., Wiley: New York, 1995. A practical guide to electrochemical techniques.

Suzuki, T.; Maruyama, Y.; Akasaka, T.; Ando, W.; Kobayashi, K.; Nagase, S. *J. Am. Chem. Soc.* **1994**, *116*, 1359. Electrochemical properties of C_{60}.

Taube, H. *Science* **1984**, *226*, 1028. Survey of electron-transfer reactions of the transition metals.

Vacuum Line Synthesis of GeH$_4$

Note: This experiment requires one 3- or 4-hour laboratory period.

Whereas the use of simple vacuum systems in synthetic inorganic chemistry dates back to the origins of chemistry, the technique as it is known today was essentially established by Alfred Stock and his co-workers in their pioneering research on the boron hydrides in the 1920s. The volatility and extreme reactivity of these hydrides (B_2H_6 spontaneously burns in air) made them particularly suited for study in a high-vacuum system. The purpose of this experiment is to introduce the most important and fundamental vacuum line techniques: transferring chemicals from one container to another on the vacuum line, measuring the vapor pressure of a liquid at low temperature, and determining the molecular weight of a gas.

As the name implies, vacuum lines are usually operated under relatively low pressures—substantially below atmospheric pressure. The unit often used for expressing low pressures is the Torr, which is defined as the pressure generated by a column of mercury 1 mm tall. Atmospheric pressure is equal to 760 Torr. The mechanical oil pump that will be used in this experiment to create the vacuum should be capable of reducing the pressure within the glass apparatus to 10^{-3} or 10^{-4} Torr. (The SI unit for pressure is the Pascal, which is defined as $1 \text{ kg m}^{-1} \text{ s}^{-2}$. Atmospheric pressure is approximately 1×10^5 Pa).

Low pressures accomplish two objectives. They remove any undesired reactive gases such as O_2 or H_2O, and they allow volatile substances to be transported rapidly from one portion of the apparatus to another. To illustrate the latter point, consider the distillation or transfer of a liquid from trap A to B in Figure 6-1. In order to make the transfer, the temperature of B must be lower than that

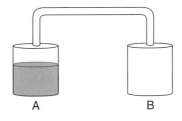

Figure 6-1
Trap-to trap transfer.

of A. Therefore, one might immerse trap B in a Dewar flask containing liquid N_2 (bp $-196\,°C$, see Appendix 9). The gaseous molecules in trap B will condense at this temperature. Molecules in trap A vaporize and diffuse by virtue of their kinetic energy to B and are condensed. The rate of diffusion of the molecules to B depends on the number of collisions that they encounter on the way to B. If the system is filled with a gas, the diffusion of molecules from A to B is very slow because of the large number of collisions that the diffusing molecules have with the gas molecules. On the other hand, if the pressure is reduced to 10^{-3} or 10^{-4} Torr (10^{-1} to 10^{-2} Pa), there are fewer molecules to get in the way, and distillation occurs more rapidly. In a vacuum, it is possible to make such transfers with any volatile compound. The volatility of a liquid is, of course, determined by its temperature; compounds having boiling points of $150\,°C$ or less at atmospheric pressure may be manipulated in a vacuum system without difficulty.

The vacuum line to be used in this experiment is shown in Figure 6-6. It is attached to a vacuum pump (see Fig. I-6 in the Introduction). The oil in the pump should be of good grade and changed regularly. After extended use, the oil accumulates chemicals handled in the vacuum line. They will increase the vapor pressure of the oil and decrease the efficiency of the pump. Because many of these chemicals are toxic, a tube should be attached to the exhaust port of the pump and vented into a hood. To protect the pump from gases, it is important that a large-volume cold trap be placed between the pump and the working portion of the vacuum line. The cold trap is cooled with a Dewar flask containing liquid N_2. While liquid N_2 is sufficiently cold ($-196\,°C$) to trap most gases used in the line, it also is cool enough to condense liquid O_2 (bp $-183\,°C$) into the trap if the trap is open to air. Because liquid O_2 is such a powerful oxidizing agent, it may react explosively with any other substances that may have also condensed into the trap. For this reason, the cold trap should never be opened to the air until the liquid N_2 trap is removed.

A vacuum line is frequently made up of several sections joined together with ground glass joints. These joints may be either of the usual standard taper type or ball-and-socket joints (Fig. 6-2); both types are used in the apparatus in Figure

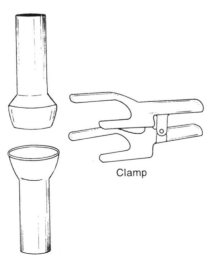

Clamp

Figure 6-2
Components of a ball-and-socket joint and clamp for holding it together.

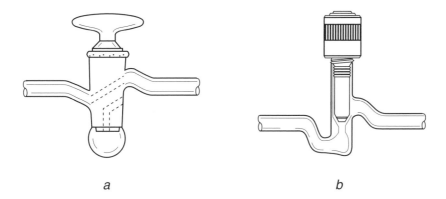

a   *b*

Figure 6-3
Glass (*a*) and greaseless (*b*) high-vacuum stopcocks. (Courtesy of Kimble/Kontes Glass Inc.)

6-6. A standard taper joint, which is usually less likely to leak, is used where bending at the joint is not required. Where some flexibility is needed, a ball-and-socket joint should be used. Both joints should be greased with either a high-vacuum silicone or Kel-F grease. The latter grease is used when the reagents would attack silicone greases. All joints should be clamped together to ensure that they do not separate and thereby allow air to leak into the vacuum line.

High-vacuum stopcocks come in two major varieties (Fig. 6-3). All-glass stopcocks, which may be evacuated, do not pop out even when the pressure in the line is somewhat above atmospheric pressure. They should be greased before being used. An ungreased stopcock should never be turned; it will become scored. The ground glass barrel and key in a vacuum stopcock are carefully matched, so it is important that these always be kept together. Because stopcocks easily break off the apparatus when put under strain, they should be turned with both hands, one to support the barrel and the other to turn the key.

It is also possible to construct a vacuum line with components that require no stopcock grease. "Greaseless" stopcocks and joints use Teflon seals and o-rings to connect adjacent parts of the vacuum line.

Although a thorough understanding of the vacuum line and its use is the most important aspect of this experiment, a few comments should be made about the reaction that will be carried out. It involves the synthesis of germane, a chemical relative of methane. It is produced by means of a multistep reaction from germanium dioxide, an inorganic polymer in which the oxygen atoms bridge between germanium centers to form a three-dimensional network. Germanium dioxide dissolves upon treatment with aqueous base to give an oxyanion written as $GeO_2(OH)^-$. This soluble source of germanium reacts with potassium tetrahydroborate (a hydride donor) to give the germyl anion GeH_3^-, which in the final step is protonated by acetic acid to give GeH_4.

$$GeO_2 + KOH \longrightarrow K[GeO_2(OH)]$$
$$K[GeO_2(OH)] + KBH_4 + H_2O \longrightarrow KGeH_3 + KB(OH)_4$$
$$KGeH_3 + HO_2C_2H_3 \longrightarrow GeH_4 + KO_2C_2H_3$$

Germane is a colorless gas at room temperature. Unlike silane, SiH_4, which spontaneously enflames in air, germane is only slowly oxidized in the air. Germane condenses to a liquid at $-88.5\,°C$ and freezes to a solid at $-165\,°C$. Its low boiling point means that liquid GeH_4 may be easily and inadvertently converted to a gas. If this occurs when the GeH_4 is confined to a small portion of the vacuum line, the pressure created by the gaseous GeH_4 will easily shatter the glass vacuum system. (If the vacuum line is designed so that it is possible to confine the gas in a small volume of the line, a safety shield placed between the worker and the line is strongly advised.)

Germane and related compounds are not just of academic interest. The hydrides and alkyl derivatives of the main group elements have many technological applications. For example, germane is used to manufacture solar cells and photoreceptors that have improved sensitivity in red and infrared (IR) light.

In the first portion of the experiment, you will generate the GeH_4 and collect it, along with other gases generated during the experiment, in a trap cooled to $-196\,°C$ with liquid N_2. You will then purify the germane by a series of trap-to-trap transfers. The less volatile impurities are removed by passing the gas mixture through a trap cooled to a temperature that condenses these impurites while allowing the germane to pass through. Highly volatile impurties can be removed by passing the gas mixture through a trap that condenses the germane but not the impurities. Gases that have volatilities similar to that of germane must be removed chemically; for example, carbon dioxide can be removed by passing the gas through a trap packed with Ascarite, which is silica coated with sodium hydroxide. Finally, you will measure the vapor pressure of germane at $-104\,°C$, determine its molecular weight, and record its gas-phase IR spectrum.

Gas-Phase Infrared Spectroscopy

Vacuum lines are most often used to study gases. To measure the IR spectrum of a gas, one uses a cell consisting of a glass tube about 2 cm in diameter and 10 cm long. A stopcock side arm is attached to the middle of the cell, and sodium chloride plates are glued to the ends of the cell (Fig. 6-4). The IR spectra of gases differ in one significant way from the spectra of solids and liquids. The IR absorptions of gases often exhibit considerable fine structure that arises from both the vibrational and rotational motions of the molecule. In contrast, the IR absorptions of solids and liquids involve only vibrational motions.

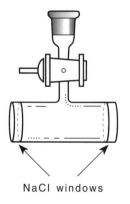

Figure 6-4
Sample cell for gas-phase IR spectroscopy.

NaCl windows

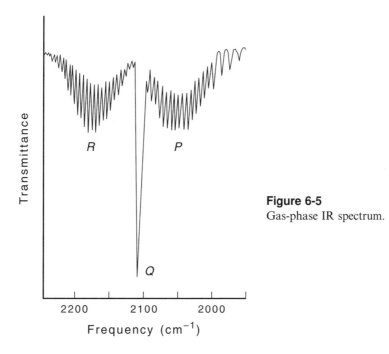

Figure 6-5
Gas-phase IR spectrum.

A typical IR band for a molecule in the gas phase is shown in Figure 6-5. The band has three components: a sharp single central absorption called the Q branch, and two sets of closely spaced bands on either side called the P and R branches. In all three branches, the same vibrational transition is involved; the branches differ in how the rotational energy of the molecule is affected.

The Q branch corresponds to a vibrational transition in which the rotational energy of the molecule is unchanged; the P and R branches correspond to vibrational transitions in which the molecule loses and gains rotational energy, respectively. The P and R branches often consist of many closely spaced lines because the rotational energy of the molecule is quantized, that is, restricted to certain special values. Each line in the gas-phase IR spectrum corresponds to a transition between two of these special energy states.

EXPERIMENTAL PROCEDURE

Safety note: The 8-h exposure limit for germane in the air is 0.2 ppm. Disposal of germane should be carried out by condensing it under vacuum into a trap at −196 °C, removing the Dewar, opening the trap to air, and allowing the trap to warm up slowly in a hood.

Before starting this experiment, read the instructions carefully. If any operation is not clear, ask for help. Although a vacuum line can be dangerous if improperly used, no difficulties should be encountered if the instructions are followed carefully. Take your time!

A serious danger has already been mentioned. Namely, at temperatures above −88.5 °C, GeH₄ is a gas; if it is allowed to vaporize in a closed section of the vacuum line, the GeH₄ pressure can shatter the vacuum line and injure persons nearby. The vacuum system shown in Figure 6-4 has relatively few locations in which the GeH_4 may be closed off from a mercury manometer. If GeH_4 vaporizes, the mercury will be forced down in the manometer and may be completely displaced from the manometer tube, thus preventing an explosion.

Germane, GeH_4

If the vacuum line is already assembled, carefully turn all of the stopcocks to determine whether they open and close easily. Care must be exercised in turning the stopcocks on the vacuum line because they can snap off, leading to a potentially dangerous situation. Turn the valves slowly, using two hands—one to hold the barrel, and one to turn the key. If any of the stopcocks do not turn easily, regrease them with silicone vacuum grease. If the vacuum line is not already assembled, clamp the necessary equipment on the supporting rack as shown in Figure 6-6. Place the clamps near the stopcocks where stress will be greatest during use. Make certain that the clamps offer firm support of the glassware but do not introduce severe strains where breakage might occur. Be sure to clamp the ball-and-socket joints.

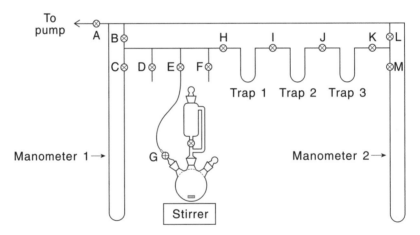

Figure 6-6
Vacuum line used for synthesis of GeH_4.

Initial configuration:

       Open stopcocks: A, B, C, H, I, J, K, L, M
       Closed stopcocks: D, E, F
       All traps at room temperature

After the components of the vacuum line are assembled, put the stopcocks in the initial configuration (see Fig. 6-6), and turn on the mechanical pump. Cool the main cold trap (to the left of valve A) to $-196\,°C$ by surrounding it with a Dewar and then filling the Dewar with liquid N_2. From time to time, check the level of liquid N_2 in the Dewar and add more if necessary.

Obtain a 500-mL three-neck round-bottom flask and equip it with a 100-mL pressure equalizing dropping funnel, an adapter for connection to a vacuum line, a stopper, and a magnetic stir bar. Grease all the ground glass joints. Make sure that the key on the dropping funnel fits well, and is lightly greased so that you can see through the hole in the key when the stopcock is turned open. Attach the flask to the vacuum line with a hose connecting stopcocks E and G, which are closed. Charge the flask with 120 mL of glacial acetic acid, and charge the dropping funnel with a solution prepared by dissolving, in order, 2 g (36 mmol) of potassium hydroxide, 1 g (9.6 mmol) of germanium dioxide, and 1.5 g (28 mmol) of potassium tetrahydroborate in 25 mL of water. Put a greased stopper in the top of the dropping funnel.

While the acetic acid is stirring, close B and open E and G. Evacuate the system through the series of traps for several minutes until the pressure stabilizes. Then close stopcock L and cool traps 2 and 3 to −196 °C with liquid N$_2$. *Note*: Only submerge the trap *bulbs* in liquid N$_2$, not the narrow "arms"; if the arms are cooled, they may clog with frozen material and make it impossible to control the pressure inside the apparatus. Slowly add the solution in the dropping funnel to the stirred acid over a period of 10–15 min. During the addition, maintain the pressure in the system (as measured by manometer 1) at approximately 100 Torr (10^4 Pa) by adjustment of stopcock L. (The pressure is 100 Torr when the mercury levels in the two arms of the manometer differ in height by 100 mm.) In this way, the pressure of hydrogen gas in the traps will probably not exceed 10 Torr (10^3 Pa), and the germane will be efficiently condensed. (*Note:* If trap 2 clogs before the addition is complete, as shown by a steadily increasing pressure in manometer 1, do the following. Stop the addition and close stopcock L. Lower the liquid N$_2$ Dewar on trap 2 and let the contents warm to room temperature and distill into trap 3. After trap 2 unclogs, reimmerse it in its liquid N$_2$ Dewar, carefully open stopcock L, and restart the addition.)

After the addition is complete, turn stopcock L so that it is fully open in order to pump most of the germane and hydrogen from the flask. After about 10 min, close stopcocks E, G, I, and L. Then disconnect the reaction vessel from the line and open it up in a hood. Combine the contents of traps 2 and 3 (germane, digermane, acetic acid, water, and carbon dioxide) in trap 3 by removing the liquid N$_2$ Dewar from trap 2 and warming the latter to room temperature.

Replace trap 1 with a trap packed with Ascarite®, and attach the calibrated bulb to the line via stopcock E. With stopcock I closed, open stopcocks B, E, and the one on the bulb. Cool trap 2 to −104 °C with a cyclohexene/liquid N$_2$ slush. (*Note*: Too much liquid N$_2$ will solidify the cyclohexene. Add liquid N$_2$ only until the solution becomes viscous.) Cool the bulb with liquid N$_2$. Close B, open I, and lower the Dewar on trap 3. Pass the germane from trap 3 through the −104 °C trap and the Ascarite trap and condense it into the bulb. After trap 3 has warmed to room temperature, it may still contain excess water and acetic acid. This material may be safely discarded. After all the volatile material has been transferred out of trap 3, close stopcocks H and I, and replace the Ascarite trap with a clean empty trap. Put fresh Parafilm® on the ends of the Ascarite trap so that the Ascarite will still be active for the next student. The contents of traps 2 and 3 should be discarded and clean traps should be mounted on the line.

The yield of germane can be determined by measuring the pressure and temperature of the gas in a system of known volume. With H and B closed, let the bulb warm to room temperature, immerse it in a water bath of known temperature, and measure the pressure on the manometer. Determine the yield of germane from your measuments and the volume of the bulb (if the volume is unknown, fill the bulb with water and weigh; determine the volume from the density of water).

The vapor pressure of germane is 310 Torr (4.12×10^4 Pa) at $-104\,°C$, and the purity of your germane can thus be checked by remeasuring the pressure with the bulb cooled with a cyclohexene/liquid N_2 slush bath. If the pressure is higher than 310 Torr, then impurities are present; if the pressure is lower than 310 Torr, then your sample of germane is too small (or the bulb is too large).

Finally, cool the bulb with liquid N_2 until the germane has recondensed. Close stopcock E, attach the IR gas cell at D, and evacuate it through stopcock B. Then close B and fill the IR cell with 10 Torr (10^3 Pa) of gas by letting the bulb warm slightly. Close stopcock D and the stopcock on the gas cell, demount the cell, and measure the infrared spectrum. When done, reattach the IR cell to the vacuum line, pump the germane into the main trap, and store the IR cell in a desiccator.

The contents of the calibrated bulb should also be pumped into the main trap.

Shut down the vacuum line as follows: (1) remove the liquid N_2 Dewar from the cold trap; (2) turn off the vacuum pump; (3) open the stopcock that vents the line to the air; and (4) remove the main cold trap and place it in the hood to allow volatile substances to evaporate. After the used cyclohexene has warmed up to near room temperature, put it back into a loosely-capped storage bottle.

REPORT

Include the following:
1. Yield of germane, with calculations.
2. Vapor pressure of germane measured at $-104\,°C$.
3. Infrared spectrum and band assignments to vibrations in GeH_4.
4. An assessment of the purity of the GeH_4 sample.

PROBLEMS

1. Why is germane more reactive than methane?
2. Which vibrational modes are IR active and which are IR inactive?
3. What is the cause of the fine structure on the IR bands?
4. What are the probable structures of $GeO_2(OH)^-$ and GeH_3^-?
5. Compare the boiling points and acidities of the Group 14 tetrahydrides. Can you explain the variations in these properties?
6. What is Ascarite®? What chemical reaction occurs when the impure gas mixture is passed over Ascarite?
7. If 2 g of GeH_4 were condensed into trap 1 at $-196\,°C$ and both adjacent stopcocks were closed, estimate the pressure (in atmospheres) that would

develop if the trap were allowed to warm up to room temperature. Assume ideal gas behavior.

8. Suppose you wanted to make a liquid N_2 slush bath that had a temperature of $-100\,°C$. How would you proceed in selecting a liquid that, when mixed with liquid N_2, would give you that temperature?

INDEPENDENT STUDIES

A. Measure the gas-phase IR spectrum of SO_2 and make band assignments. (Briggs, A. G. *J. Chem. Educ.* **1970**, *47*, 391.)

B. Prepare and characterize $BF_3:N(CH_3)_3$. (Amster, R. L.; Taylor, R. C. *Spectrochim. Acta* **1964**, *20*, 1487. Bryan, P. S.; Kuczkowski, R. L. *Inorg. Chem.* **1971**, *10*, 200. Burg, A. B.; Green, A. A. *J. Am. Chem. Soc.* **1943**, *65*, 1838. Hartman, J. S.; Miller, J. M. *Inorg. Chem.* **1974**, *13*, 1467. Katritzky, A. R. *J. Chem. Soc.* **1959**, 2049.)

C. Make $CF_2=CF_2$ by depolymerization of Teflon tape (Hunadi, R. J. *Synthesis* **1982**, 454), and study its reactions with Vaska's complex or nickelocene.

D. Prepare $B_3N_3H_6$ on a vacuum line. (Wideman, T.; Sneddon, L. G. *Inorg. Chem.* **1995**, *34*, 1002.)

REFERENCES

GeH₄ and Related Main Group Hydrides
Das, P. P.; Devi, V. M.; Rao, K. N. *J. Mol. Spec.* **1982**, *91*, 494. Infrared spectrum of germane.
Jolly, W. L.; Drake, J. E. *Inorg. Synth.* **1963**, *7*, 34. Synthesis of germane.
Rochow, E. G. in *Comprehensive Inorganic Chemistry*, Bailar, J. C. Jr.; Emeleus, H. J.; Nyholm, R.; Trotman-Dickenson, A. P., Eds, Pergamon: Oxford, UK, 1973; Vol. 2, pp 1–41. Review of germanium chemistry.
Stone, F. G. A. *Hydrogen Compounds of the Group IV Elements*, Prentice Hall: New York, 1962.
Todd, M.; McMurran, J.; Kouvetakis, J.; Smith, D. *J. Chem. Mater.* **1996**, *8*, 2491. Chemical vapor deposition from germanium hydrides.

Vacuum Techniques
Errington, R. J., Ed. *Advanced Practical Inorganic and Metalorganic Chemistry*, Blackie: New York, 1997.
Shriver, D. F.; Drezdzon, M. A. *The Manipulation of Air-sensitive Compounds*, 2nd ed., Wiley: New York, 1986. The construction and use of vacuum lines.

Vibrational and Rotational Spectroscopy
Nakamoto, K. *Infrared and Raman Spectra of Inorganic and Coordination Compounds*, 5th ed., Wiley: New York, 1997. A two-volume compendium of the theory and practice of IR and Raman spectroscopy.

Tin Chemistry: Coordination Complexes and Organometallic Derivatives

Note: Part A of this experiment requires 6 hours and Part B requires 2 hours.

As a member of the Group 14 elements, tin forms many compounds that are analogues of known carbon compounds. For example, $SnCl_4$ and CCl_4 are both colorless liquids, and their boiling points are very similar: 114 and 77 °C, respectively. In general, however, the chemistries of tin and carbon differ considerably. These differences arise from two factors: Sn atoms are larger than C atoms, and Sn is less electronegative than C. Both of these factors favor the tendency of Sn to bond to more than four atoms. For example, $SnCl_4$ reacts readily with various donor molecules, L, to form six-coordinate adducts:

$$SnCl_4 + 2L \longrightarrow SnCl_4L_2$$

$$L = N(CH_3)_3, O(C_2H_5)_2, \text{ or } S(C_2H_5)_2$$

$$SnCl_4 + 2Cl^- \longrightarrow SnCl_6{}^{2-}$$

These adducts are octahedral and the $SnCl_4L_2$ complexes usually adopt trans geometries.

In this experiment, you will prepare an adduct of $SnCl_4$ with dimethyl sulfoxide, $OS(CH_3)_2$, which is commonly known as DMSO:

$$SnCl_4 + 2OS(CH_3)_2 \longrightarrow SnCl_4[OS(CH_3)_2]_2$$

Infrared (IR) studies suggest that the product has a trans geometry. Dimethyl sulfoxide exhibits a phenomenon called *linkage isomerism*: it coordinates to some metals through the S atom, and to others through the O atom. In order to distinguish these coordination modes, it is useful to consider the two major resonance structures of DMSO.

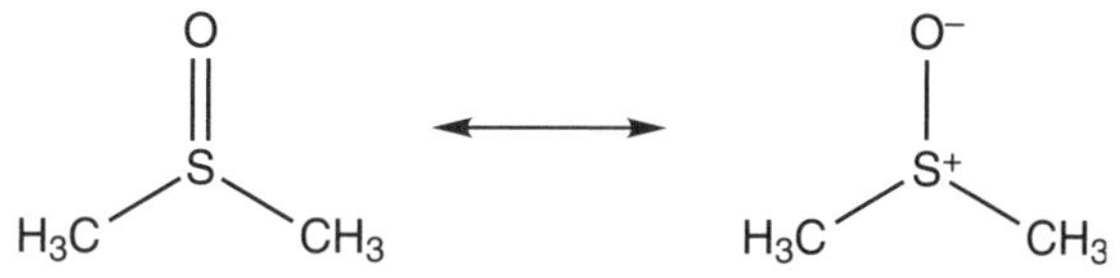

When coordination occurs at oxygen, the second resonance structure is dominant because it places more negative charge on oxygen. In this bonding mode, the S–O bond is weakened and the frequency of the S–O vibration is lowered. On the other hand, coordination at sulfur will have the opposite effect, leading to a stronger S–O bond and a higher frequency S–O vibration. In this experiment, you will establish the coordination mode of the DMSO ligand in $SnCl_4[OS(CH_3)_2]_2$ by examining its IR spectrum.

In another part of this experiment, you will explore some aspects of the organometallic chemistry of tin. For example, $SnCl_4$ reacts with Grignard reagents to give organotin compounds:

$$SnCl_4 + 4\,RMgX \longrightarrow SnR_4 + 4\,MgXCl$$

Such alkylation reactions are important in the chemistry of the main group elements, and similar reactions are widely used in the preparation of SiR_4, GeR_4, BR_3, AlR_3, PR_3, AsR_3, and SbR_3 compounds from the corresponding chloro starting materials. In most cases, organolithium compounds, RLi, also can be used instead of the Grignard reagents.

Organotin compounds such as $SnClR_3$ and $SnCl_2R_2$ are produced on a large scale industrially. In principle, it is possible to synthesize the mixed compounds $SnCl_xR_{4-x}$ by treating $SnCl_4$ with fewer than four equivalents of RMgX. In practice, such reactions afford mixtures of compounds that are difficult to separate. The best method to prepare them is frequently the "redistribution" reaction of stoichiometric amounts of $SnCl_4$ and SnR_4:

$$SnCl_4 + 3\,SnR_4 \longrightarrow 4\,SnClR_3$$

Dialkyltin thiolates and carboxylates are used as stabilizers in plastics such as polyvinylchloride (PVC), and trialkyltin carboxylates are used as biocidal coatings on ship hulls. Other organotin compounds serve as disinfectants or as fungicides. Organotin compounds are also widely used in organic synthesis. Some of the most important organotin reagents are the hydrides, which are prepared by the reaction of $SnCl_xR_{4-x}$ with $LiAlH_4$ to give SnH_xR_{4-x}. The versatile chemistry of organotin reagents is illustrated by the synthesis of the interesting compound arsabenzene:

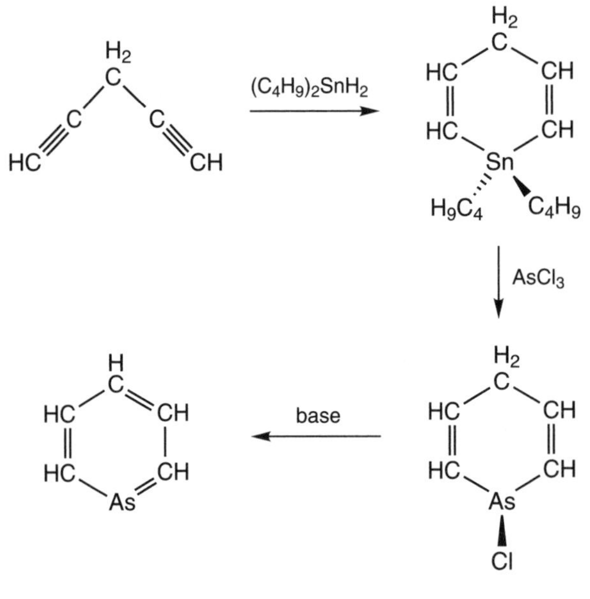

In the present experiment, the specific reaction that will be carried out is:

$$2\,Sn\; +\; 3\,ClCH_2C_6H_5\; \longrightarrow\; SnCl(CH_2C_6H_5)_3\; +\; SnCl_2$$

The product, chlorotribenzyltin, is a colorless solid that is stable in air for several days. In some respects, the reaction is analogous to the preparation of Grignard reagents, $RMgX$, by treatment of magnesium metal with organic halides. One difference is that the reaction of tin metal with benzyl chloride can be carried out in water, whereas Grignard reagents react instantly with water and must be prepared in aprotic solvents such as diethyl ether.

Unlike $SnCl_4$, chlorotribenzyltin does not form stable six-coordinate complexes with donor molecules. In general, the tendency to form such complexes decreases in the order $SnCl_4 > SnCl_3R > SnCl_2R_2 > SnClR_3 > SnR_4$.

Because the Sn–C and Sn–Cl vibrations occur at frequencies below 650 cm^{-1}, the IR spectrum of $SnCl(CH_2C_6H_5)_3$ simply consists of $CH_2C_6H_5$ vibrational absorptions that reveal very little about the composition or structure of the compound. The mass spectrum of $SnCl(CH_2C_6H_5)_3$ contains much more information about the structure of the compound. For a discussion of mass spectrometry, see Experiment 18. The large number of stable isotopes of Sn aids in the assignment of peaks to ion fragments containing Sn, but also can make the spectra more difficult to interpret. The isotopes of Sn and their natural abundances are

Isotope	Natural Abundance (%)
^{112}Sn	0.95
^{114}Sn	0.65
^{115}Sn	0.34
^{116}Sn	14.2
^{117}Sn	7.6
^{118}Sn	24.0
^{119}Sn	8.6
^{120}Sn	33.0
^{121}Sn	4.7
^{124}Sn	6.0

Another technique that can also provide information about the structures of many compounds is nuclear magnetic resonance spectroscopy.

Nuclear Magnetic Resonance Spectroscopy

Electrons behave as if they were spinning spheres of negative charge. Experimental measurements suggest that electrons may spin in clockwise or counterclockwise directions around an axis to generate a magnetic field, ↑ or ↓. In the same way, many nuclei, such as ^1H, ^{31}P, and ^{19}F, spin in either of two ways to generate ↑ or ↓ magnetic fields. Whereas nuclei of opposite spins normally have the same energies, in an externally applied magnetic field this is no longer true.

Then the field produced by the spinning nucleus is either aligned with the applied field or opposes it. A nucleus aligned (↑) with the applied field has a lower energy than one opposed (↓) to it.

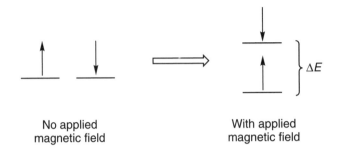

No applied
magnetic field

With applied
magnetic field

These spin states are separated by a relatively small energy ΔE. By irradiating a nucleus with radio waves, it is possible to cause the nucleus to go from the low energy state (↑) to the high energy state (↓) with concomitant absorption of a radio frequency having the energy ΔE. If the nucleus is in the high energy state, it can also return to the low energy state by emitting radio waves with the same energy ΔE. This absorption or emission of radio waves is measured to give the nuclear magnetic resonance (NMR) spectrum. The magnitude of ΔE depends on the strength of the applied magnetic field, the spin properties of the nucleus under study, and the extent to which the spinning nucleus is shielded from the applied magnetic field by electrons surrounding the nucleus.

As noted in Experiment 12, electrons can create a magnetic field that opposes any applied field. Thus, electron density, which is determined by the chemical environment of the nucleus in a molecule, is a major factor influencing the value of ΔE. For this reason, chemically different ^1H atoms in a molecule will have different transition energies ΔE. For example, in $SnCl(CH_2C_6H_5)_3$, there are four types of ^1H nuclei because there are four chemically different types of hydrogen atoms. The ↑ to ↓ transition will occur at different ΔE values for each kind of ^1H nucleus. Thus, an NMR spectrum of $SnCl(CH_2C_6H_5)_3$ will show absorption of four energies of radio frequency radiation. Moreover, the amount of absorbed radiation depends on the number of each kind of H atom. For $SnCl(CH_2C_6H_5)_3$, the ratio of the amount of absorbed radio waves for the CH_2 and the ortho, meta, and para protons of the benzyl group is 2:2:2:1, respectively. The NMR spectrum of $SnCl(CH_2C_6H_5)_3$ shown in Figure 7-1 contains four features; the integrated areas of the features are in the expected ratio. Thus, NMR spectroscopy allows one to count the number of hydrogen atoms that are chemically different in a molecule.

The transition energy, ΔE, is usually expressed as a radio frequency, cycles per second (hertz, Hz, is commonly used for cps). Rather than simply specifying the absolute frequency, ν_{samp}, required for the ↑ to ↓ transition for a given proton in the sample, chemists report instead the *difference* between ν_{samp} and the resonance frequency of some reference compound, ν_{ref}. This difference divided by the frequency of the reference is a unitless number called the chemical shift:

$$\text{chemical shift} = \frac{\nu_{samp} - \nu_{ref}}{\nu_{ref}}$$

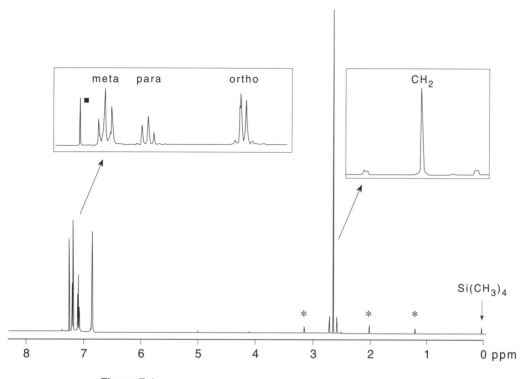

Figure 7-1
The ^1H NMR spectrum of $SnCl(CH_2C_6H_5)_3$.
Key: $*$ = impurities, ■ = chloroform solvent resonance.

The reference compound for ^1H NMR spectroscopy is tetramethylsilane (TMS), $Si(CH_3)_4$. The equation above thus fixes the chemical shift of TMS at 0. For most ^1H NMR spectra, the peaks are found at chemical shifts between 0 and 8×10^{-6}. To avoid using the 10^{-6} factor, chemical shifts are usually expressed as parts per million (ppm), in which case the symbol δ is used: for example, δ 8.0.

In the above discussion, every different kind of H atom in a molecule was assumed to absorb radio waves of only one frequency. Owing to a very important phenomenon called spin-spin coupling, however, it is possible for a particular kind of H atom to absorb several different frequencies of radio waves. The spectrum of $SnCl(CH_2C_6H_5)_3$ shown in Figure 7-1 shows this phenomenon. For example, the ortho protons absorb radio waves of two different energies and give rise to a two-line pattern in the NMR spectrum called a doublet. In addition, the meta and para protons absorb radio waves of three different energies and give rise to three-line patterns called triplets. Furthermore, the CH_2 protons give rise to a more complicated five-line pattern in the NMR spectrum.

The multiple lines seen in Figure 7-1 for each kind of H atom arise from magnetic interactions between the nuclei of nearby atoms in the molecule. Consider the nucleus of one of the para protons in $SnCl(CH_2C_6H_5)_3$. If this nucleus did not engage in spin-spin coupling, then it would absorb only one frequency of radio waves and it would appear as a singlet in the NMR spectrum. In fact, however, the para H atom is situated close to the two meta protons. The nucleus of each meta proton generates a small magnetic field that depends on whether the meta nucleus is spin up or spin down. Consequently, the magnetic field experienced by the nucleus under consideration, the para nucleus in $SnCl(CH_2C_6H_5)_3$, will be one value if both meta neighbors are spin up, another value if both are spin down, and a third value if one meta neighbor is spin up and the other spin down. As a result, the para nucleus will absorb radio waves of three different frequencies.

Interestingly, the five-line pattern seen for the CH_2 proton is not due to spin-spin coupling between H atoms; instead these two hydrogen atoms are coupled to the Sn nucleus. Two of the small outer lines are due to molecules in which the ^{117}Sn isotope is present, the other two small outer lines are due to molecules in which the ^{119}Sn isotope is present, and the large central line is due to molecules in which one of the other isotopes of Sn is present (none of which is able to engage in spin-spin coupling—see Appendix 7). Thus the observed spectrum of $SnCl(CH_2C_6H_5)_3$ is actually the sum of three spectra: the spectrum of $^{119}SnCl(CH_2C_6H_5)_3$ molecules, the spectrum of $^{117}SnCl(CH_2C_6H_5)_3$, and the spectrum of $SnCl(CH_2C_6H_5)_3$ molecules containing the other isotopes of Sn.

The purpose of this exceedingly brief treatment of NMR has been to relate the parameters obtained from an NMR spectrum to the origins of the NMR phenomenon. The discussion has greatly oversimplified the theoretical background upon which NMR spectroscopy is based. Finally, we have not mentioned the different types of NMR instruments: continuous wave (CW) and Fourier transform (FT). You are encouraged to examine the references listed at the end of this experiment for further information about NMR spectroscopy.

EXPERIMENTAL PROCEDURE

Part A

Chlorotribenzyltin, $SnCl(CH_2C_6H_5)_3$

Safety note: Benzyl chloride is poisonous as well as a skin and eye irritant; it should be handled in a hood with proper protection. The toxicity of organotin compounds depends on the number and type of organic groups bound to the tin center. Methyl and ethyl derivatives can be quite toxic. Benzyl tin compounds are not known to be toxic, but they should be treated with caution.

The tin powder and benzyl chloride should be fresh; aged samples will give poor yields.

In a hood, place 2.0 g (17 mmol) of 325 mesh tin powder, 4.0 mL of H_2O, and 6.0 mL (6.6 g, 52 mmol) of benzyl chloride in a 50-mL one-neck round-bottom flask. Attach a water-cooled condenser to the flask, using silicone grease to lubricate the joint. Place a rheostat-controlled heating mantle under the flask, turn on the cooling water, and use a voltage regulator to heat the reaction mixture to reflux for 3 h (see Fig. 7-2 for a diagram of the apparatus). Set the voltage on the rheostat high enough so that the boiling water heats the greased joint sig-

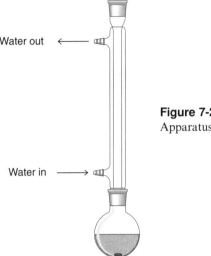

Water out ←

Water in →

Figure 7-2
Apparatus used to prepare $SnCl(CH_2C_6H_5)_3$.

nificantly above room temperature. (Do not try to wash any solids that may collect on the inside walls of the flask back into the solution; your attempts to do so will only make things worse.)

After the reaction mixture has been heated to reflux for 3 h,* remove the heat and cool the flask in ice. The lower liquid phase should solidify to form a white solid. When the reaction flask is cool to the touch, pour off the liquid from the solid mass (*note*: the liquid is the lower layer). *Here is a good stopping point for the first day's work.* Stopper the flask containing the white solid until the next period. You may discard the liquid.

Extract the white solid with 10 mL of acetone, and suction-filter the extract (see Figure 13-1). Repeat the extraction with an additional 10 mL of acetone. Discard the gray powder (unreacted tin). Place the filtered acetone extracts in a 50-mL one-neck round-bottom flask, and use a rotary evaporator to remove the acetone under reduced pressure (if there is a water bath on the rotary evaporator, it can be warmed to speed the evaporation). A slightly wet-looking white solid should remain. To recrystallize your product, add a stirring bar and 8 mL of ethyl acetate to the flask. In a hood, heat the mixture to boiling with a rheostat-controlled heating mantle. The white solid should dissolve, but if it does not, add an extra 2 mL of ethylacetate and continue heating until no solid remains. As soon as the solid has dissolved, use insulated gloves or a clamp to transfer the flask to an ice bath. Cool the mixture for 30 min, and then collect the solid using a suction filtration apparatus (Fig. 13-1). Additional product can be obtained by concentrating the filtrate further, and cooling it in ice.

---

*The white solids that appeared shortly after refluxing commenced should have largely disappeared, although small amounts of white or gray solids on the sides of the flask are usually still present. The reaction mixture at this point should consist of two liquid phases, plus a small amount of unreacted tin.

Weigh your product. For long-term storage, the product should be kept under an N_2 atmosphere.

Record the NMR spectrum of your product (see next section). Also, analyze the mass spectrum given in Appendix 10. Using the isotopic abundances listed earlier, assign all of the peaks in the spectrum in a manner similar to that outlined in Experiment 18. [The naturally occurring isotopes of Cl and their abundances are ^{35}Cl (75.4%) and ^{37}Cl (24.6%); see Appendix 7.]

Solution NMR Spectrum

Almost all commercially available NMR instruments require liquid samples of approximately 1-mL volume. Because many samples are solid, they must first be dissolved in a solvent that does not absorb in the region of the NMR spectrum of interest.

Dissolve about 50 mg of $SnCl(CH_2C_6H_5)_3$ in 1 mL of chloroform (or deuterated chloroform; see your instructor for information about preparing the NMR sample.) With a pipet, transfer the solution to a 5 mm diameter NMR tube. The solution should fill the NMR tube to a height of between 4 and 7 cm. If you are unable to record the NMR spectrum yourself, interpret the spectrum in Figure 7-1. If you are able to record the NMR spectrum, consult your instructor for information about how to use the instrument.

The ^1H NMR spectrum contains signals from the CH_2 protons, and the ortho, meta, and para protons of the phenyl ring. The protons on the phenyl ring give rise to a set of resonances near δ 7. The phenyl resonances are located near the resonance due to the chloroform solvent, which appears as a single sharp peak at δ 7.24. The ^1H NMR resonance of the CH_2 group shows "satellites," which are smaller peaks that appear on both sides of the main resonance. These satellites are due to the presence of the isotopes ^{117}Sn and ^{119}Sn (see Appendix 7), which have nuclear spins of $\frac{1}{2}$, and which can couple with nearby ^1H nuclei. The coupling constant between the tin and hydrogen nuclei is given the symbol $J(Sn–H)$, and is reported in hertz. The size of the coupling constant is the frequency difference between the satellite peaks.

Part B

Tetrachlorobis(dimethyl sulfoxide)tin (IV), $SnCl_4[OS(CH_3)_2]_2$

Safety note: Dimethyl sulfoxide has the unusual and potentially dangerous property of readily passing through the skin. Whereas DMSO itself is not very toxic, it is hazardous because of its ability to transport toxic solutes into the circulatory system. Tin tetrachloride is irritating to the eyes.

In a hood, dissolve 2.25 mL (5 g, 19 mmol) of $SnCl_4$ in 45 mL of anhydrous diethyl ether in a 125-mL Erlenmyer flask. Then add a solution of 2.9 mL (3.2 g, 41 mmol) of DMSO in 5 mL of ether. Upon standing, the solution deposits a white precipitate of $SnCl_4[OS(CH_3)_2]_2$. Suction filter, and dry the precipitate on the fritted glass filter. Calculate the percentage yield and determine the melting point of the product. Record its IR spectrum as a Nujol mull. The S–O stretching mode of uncoordinated $OS(CH_3)_2$ occurs at approximately 1100 cm^{-1}. Try to locate the S–O absorption in the spectrum of your sample. From the change in frequency from free $OS(CH_3)_2$, suggest whether $OS(CH_3)_2$ is coordinated to Sn through the S or O atom.

REPORT

Include the following:
1. Yield and melting point of $SnCl(CH_2C_6H_5)_3$.
2. Assignment and interpretation of the ^1H NMR spectrum of $SnCl(CH_2C_6H_5)_3$.
3. Analysis of the mass spectrum of $SnCl(CH_2C_6H_5)_3$.
4. Yield, melting point, and IR spectrum of $SnCl_4[OS(CH_3)_2]_2$.
5. Comparison of IR spectra of $OS(CH_3)_2$ and $SnCl_4[OS(CH_3)_2]_2$.
6. Assignment of the $OS(CH_3)_2$ bonding mode in $SnCl_4[OS(CH_3)_2]_2$.

PROBLEMS

1. Why are the NMR chemical shifts of the $Sn–CH_2$ protons and the C_6H_5 protons of $SnCl(CH_2C_6H_5)_3$ so different?
2. Propose a chemical formula for the white solid that forms during the initial stages of the reaction of Sn with benzyl chloride.
3. As judged from reactions that you carried out in this experiment, is $O(C_2H_5)_2$ or $OS(CH_3)_2$ the stronger Lewis base toward $SnCl_4$?
4. Draw all possible structural and optical (if any) isomers of $SnCl_2Br_2[OS(CH_3)_2]_2$.
5. Explain the following trend in the tendency for $SnCl_xR_{4-x}$ compounds, where R = alkyl, to coordinate additional ligands: $SnCl_4 > SnCl_3R > SnCl_2R_2 > SnClR_3 > SnR_4$.
6. Suggest a method for preparing $P(C_2H_5)_3$.

INDEPENDENT STUDIES

A. Prepare and characterize $SnCl_3(C_4H_9)$ from $SnCl_4$ and $Sn(C_4H_9)_4$. (Neuman, W. P.; Burkhardt, G. *Justus Liebigs Ann. Chem.* **1963**, *663*, 11.)
B. Prepare and characterize the pyridine (py) adduct $SnCl_4(py)_2$. (Laubengayer, A. W.; Smith, W. C. *J. Am. Chem. Soc.* **1954**, *76*, 5985.)
C. Prepare and characterize triphenyltin hydride, $SnH(C_6H_5)_3$. (Allen, C. W. *J. Chem. Educ.* **1970**, *47*, 479.)
D. Prepare and characterize $SnCl_2(acetylacetonate)_2$ obtained from $SnCl_4$ and acetylacetone. (Thompson, D. W.; Kranbuehl, D. E.; Schiavelli, M. D. *J. Chem. Educ.* **1972**, *49*, 569.)

REFERENCES

$SnCl_4[OS(CH_3)_2]_2$ and $SnCl(CH_2C_6H_5)_3$
Chambers, D. B.; Glockling, F.; Weston, M. *J. Chem. Soc. (A)* **1967**, 1759. Mass spectrometry of organotin compounds.

Sisido, K.; Kozima, S.; Hanada, T. *J. Organomet. Chem.* **1967**, *9*, 99. Mechanism of the reaction of tin with benzyl chloride.

Sisido, K.; Takeda, Y.; Dinugawa, Z. *J. Am. Chem. Soc.* **1961**, *83*, 538. Preparation of chlorotribenzyltin.

Tanaka, T. *Inorg. Chim. Acta* **1967**, *1*, 217. Preparation and IR spectra of DMSO adducts of $SnCl_4$ and $SnCl_2R_2$.

Verdonck, L.; van der Kelen, G. *J. Organomet. Chem.* **1966**, *5*, 532. The NMR spectrum of chlorotribenzyltin.

Organotin Chemistry

Ashe, A. A. III *Acc. Chem. Res.* **1978**, *11*, 153. Syntheses using tin heterocycles.

Davies, A. G. *Organotin Chemistry*, VCH: New York, 1997.

Eisch, J. J. *Organometallic Syntheses*, Academic: New York, 1981, Vol. 2.

Elschenbroich, Ch.; Salzer, A. *Organometallics, A Concise Introduction*, 2nd ed., VCH: New York, 1992.

Evans, C. J.; Karpel, S., Eds. *Organotin Compounds in Modern Technology*, Elsevier: Amsterdam, 1985. Applications and toxicology of organotin compounds.

Molloy, K. C. *Adv. Organomet. Chem.* **1991**, *33*, 171. A review of cyclic organotin compounds.

Patai, S., Ed. *The Chemistry of Organic Germanium, Tin, and Lead Compounds*, Wiley: New York, 1995.

Sekingushi, A.; Sakurai, H. *Adv. Organomet. Chem.* **1995**, *37*, 1. A review of cage compounds containing Si, Ge, and Sn.

Sita, L. *Adv. Organomet. Chem.* **1995**, *38*, 189. Chemistry of polystannanes.

NMR Spectroscopy

http://bmrl.med.uiuc.edu:8080/MRITable/. A listing of the NMR properties of the elements.

Derome, A. E. *Modern NMR Techniques for Chemistry Research*, Pergamon: New York, 1987. Introduction to the theory and interpretation of NMR spectra.

Drago, R. S. *Physical Methods for Chemists*, Saunders: Orlando, FL, 1992, Chapter 7. Basic treatment of NMR spectroscopy, with applications to inorganic compounds.

Friebolin, H. *Basic One- and Two-Dimensional NMR Spectroscopy*, 3nd ed., VCH: Weinheim, 1998.

Mason, J., Ed. *Multinuclear NMR*, Plenum: New York, 1987.

Silverstein, R. M.; Bassler, G. C.; Morrill, T. C. *Spectrometric Identification of Organic Compounds*, 5th ed.; Wiley, New York, 1991.

IR Spectra and Chemistry of DMSO Compounds

Goggin, P. L. in *Comprehensive Coordination Chemistry*, Wilkinson, G.; Gillard, R. D.; McCleverty, J. A., Eds., Pergamon: New York, 1987; Vol. 2, Chap. 15.8.

Nakamoto, K. *Infrared Spectra of Inorganic and Coordination Compounds*, 5th ed., Wiley: New York, 1997. Infrared spectra of DMSO complexes.

Synthesis of $(C_6H_5)_2PCH_2CH_2P(C_6H_5)_2$ in Liquid Ammonia

Note: This experiment requires 8 hours spread over two laboratory periods.

To a chemist, water is far from an ordinary liquid. For example, its heat capacity, dielectric constant, and boiling point are all much higher than expected for such a small molecule. The unusual properties of water all reflect the presence of strong hydrogen bonds between the individual molecules. It is these unusual properties (and some others) that make water uniquely suited as a basis for life on earth. What other substance most closely resembles water? To the chemist there in one obvious answer: ammonia. Here we are not referring to the ammonia that is used as a household cleaner, which is actually a mixture of ammonia and water. Pure ammonia is a gas a room temperature, but it can be liquified by cooling it to $-33\,°C$. Like water, liquid ammonia exhibits hydrogen-bonding interactions and it has many unusual properties as a consequence.

Owing to its useful properties, ammonia is widely used in chemical synthesis both as a reactant and as a solvent. With its low boiling point, of course, it will vaporize at a significant rate if it is used without external cooling. This problem is not serious, however, if the experiments are conducted in an efficient hood. In certain cases, any inconvenience created by the low boiling point of the liquid and the noxious nature of the gas is greatly exceeded by the advantageous chemical properties of the solvent. For example, anhydrous liquid NH_3 has the very unusual property of dissolving certain highly reducing metals. Thus, ammonia dissolves the alkali metals, the alkaline earths (except Be), the lanthanide metals Eu and Yb (which form divalent cations), and even some metals such as Y that form trivalent cations. Numerous studies of the electrical conductivity as well as optical and magnetic properties of such solutions indicate that they contain solvated (i.e., ammoniated) metal cations and solvated electrons:

$$M + NH_3 \longrightarrow M(NH_3)_x^{+} + e(NH_3)_y^{-}$$

Usually, solvated electrons quickly react with solvents and so are destroyed. In ammonia and closely related amines, however, these solvated electrons are stable for long periods. Metal–ammonia solutions are characteristically blue although at higher concentrations a bronze color is observed. The color is a consequence of the solvated electrons. Evaporation of these solutions affords mirrors of the metal.

To be sure, alkali metals do react with NH_3, but very slowly. The reaction is analogous to that of such metals with H_2O:

$$2\,Na\;+\;2\,NH_3\;\longrightarrow\;2\,NaNH_2\;+\;H_2$$

The transformation is marked by the disappearance of the intense blue color of the solvated electrons. Iron salts (even rust!) catalyze this reaction and care must be taken to prevent the accidental addition of iron compounds unless one wishes to prepare MNH_2 compounds.

Sodium amide, $NaNH_2$, in NH_3 is analogous to sodium hydroxide, $NaOH$, in H_2O. As OH^- is called a base in H_2O, NH_2^- is called a base in NH_3. Like H_2O, NH_3 dissociates or autoionizes but to a significantly lesser extent:

$$2\,H_2O\;\rightleftharpoons\;H_3O^+\;+\;OH^-\qquad K_{water}\;=\;1.0\;\times\;10^{-14}\quad\text{at }25\,°C$$

$$2\,NH_3\;\rightleftharpoons\;NH_4^+\;+\;NH_2^-\qquad K_{ammonia}\;=\;1.9\;\times\;10^{-33}\quad\text{at }-50\,°C$$

Thus, basic NH_3 solutions are those that contain a higher concentration of NH_2^- than of NH_4^+, and acidic solutions are those that contain a higher concentration of NH_4^+ than of NH_2^-. The NH_4^+ ion is not a very strong proton donor compared to H_3O^+ in H_2O; thus, for reactions requiring a strong acid or proton donor, liquid NH_3 is a poor choice. On the other hand, a reaction that requires a very strong base will greatly benefit from the higher basicity of NH_2^- compared with OH^-.

Solutions of alkali metals in ammonia are very strong reducing agents. These solutions will reduce aromatic hydrocarbons, many main group elements, and metal complexes. For example, $Ni(CN)_4^{2-}$ can be reduced with potassium metal in liquid NH_3 to give $Ni(CN)_4^{4-}$, an ion that is isoelectronic with $Ni(CO)_4$ and that contains Ni in an oxidation state of 0:

$$Ni(CN)_4^{2-}\;+\;2\,e(NH_3)_y^{\,-}\;\longrightarrow\;Ni(CN)_4^{4-}\;+\;2y\,NH_3$$

In this reaction, the liquid NH_3 not only provides a medium for accommodating the very reactive metallic potassium but also is a sufficiently polar solvent to dissolve the reacting metal complex. The product of the reaction, $K_4[Ni(CN)_4]$, is obtained simply by evaporation of the ammonia.

In this experiment, you will use liquid ammonia to prepare a diphosphine that can potentially serve as a bidentate ligand for transition metals. In the first step of the synthesis, you will take advantage of the strongly reducing nature of sodium–ammonia solutions to add two electrons to triphenylphosphine, which results in cleavage of a carbon-phosphorus bond. One of the products of this reaction, phenylsodium (NaC_6H_5), is the conjugate base of benzene. Owing to its high basicity, NaC_6H_5 quickly deprotonates ammonia to give benzene and $NaNH_2$. You will neutralize the resulting solution of $NaNH_2$ by adding NH_4Br.

The other product of the cleavage of $P(C_6H_5)_3$ with the Na/NH_3 solution is $NaP(C_6H_5)_2$, sodium diphenylphosphide. Sodium diphenylphosphide is a yellow salt that is soluble in polar solvents such as ammonia. Because it is very easily oxidized, $NaP(C_6H_5)_2$ must be protected from air. The final step in this procedure is the reaction of $NaP(C_6H_5)_2$ with 1,2-dichloroethane to give 1,2-bis(diphenylphosphino)ethane, $(C_6H_5)_2PCH_2CH_2(C_6H_5)_2$.

$$P(C_6H_5)_3 + 2\,Na \longrightarrow NaP(C_6H_5)_2 + NaC_6H_5$$

$$NaC_6H_5 + NH_3 \longrightarrow NaNH_2 + C_6H_6$$

$$NaNH_2 + NH_4Br \longrightarrow NaBr + 2\,NH_3$$

$$2\,NaP(C_6H_5)_2 + ClCH_2CH_2Cl \longrightarrow (C_6H_5)_2PCH_2CH_2P(C_6H_5)_2 + 2\,NaCl$$

The diphosphine $(C_6H_5)_2PCH_2CH_2P(C_6H_5)_2$, which is usually abbreviated dppe, is widely used as a bidentate ligand for transition metals. In this experiment, you will prepare the nickel complex $NiCl_2(dppe)$:

$$Ni(H_2O)_6Cl_2 + dppe \longrightarrow NiCl_2(dppe) + 6\,H_2O$$

The complex $NiCl_2(dppe)$ is used as a catalyst for the cross-coupling of aryl halides and Grignard reagents:

$$Ar-Br + Ar'-MgBr \xrightarrow{\;NiCl_2(dppe)\;} Ar-Ar' + MgBr_2$$

EXPERIMENTAL PROCEDURE

Safety note: Liquid ammonia is a very volatile, irritating, and toxic material; all operations involving it must be conducted in a hood. Wear rubber gloves whenever handling liquid NH_3. Although ammonia has a low boiling point ($-33\,°C$), its heat of vaporization is sufficiently high that it can be dispensed into the reaction flask without external cooling. Solutions of sodium in liquid ammonia are so reducing that they carbonize Teflon-coated stir bars; it is therefore necessary to use a glass-covered stir bar. The large amount of NH_3 gas evolved during the operation serves as a protective atmosphere so it is not necessary to flush the reaction flask with N_2.

Sodium–Liquid NH_3 Solution

Equip a 500-mL three-neck round-bottom flask with a glass covered magnetic stir bar (*Caution*: Rapidly agitated glass stir bars tend to break through round-bottom flasks). Attach a dry ice condenser to the central neck, attach the condenser to a bubbler (see Fig. 8-1), but leave one neck unstopped for now. Lubricate all joints with silicone grease.

The next step is to put liquid NH_3 into the flask. Ammonia cylinders contain a mixture of liquid and gas, and which of these comes out of the cylinder depends on how the cylinder is oriented. The larger NH_3 cylinders usually are fitted with a

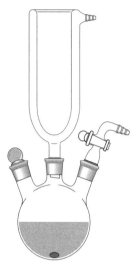

Figure 8-1
Three-neck flask equipped with a dry ice condenser.

goose neck eductor tube as shown in Figure 8-2. With the cylinder tipped and the outlet of the main valve pointing upward, the eductor tube will be immersed in the liquid NH_3. (Smaller cylinders without eductor tubes may simply be tipped at a greater angle.) In this way, liquid NH_3 is dispensed from the cylinder directly into the reaction flask. This procedure is by far the best way to obtain the 200 mL needed for the reaction. (Do *not* try to obtain the liquid NH_3 by cooling the gas with a dry ice condenser. This latter method will condense enough water from the air to spoil the reactions in the next step.)

Delivery of the liquid NH_3 from the cylinder may be accomplished through a rubber tube inserted into the open neck of the flask. After positioning the cylinder as shown in Figure 8-2, open the main valve (*Remember*, this is only an on–off valve). Regulate the flow of liquid NH_3 by adjusting the needle valve. Introduce approximately 200 mL of liquid NH_3 into the flask. After the NH_3 has

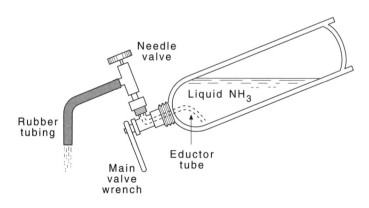

Figure 8-2
Dispensing NH_3 from a cylinder.

been added to the reaction flask, close the needle valve and main valve on the NH_3 cylinder, and stopper the third neck of the flask. The ammonia that remains in the transfer tube should be allowed to evaporate in a hood.

Charge the condenser to *one-quarter* full with 95% ethanol and then cautiously add pieces of dry ice. The dry ice condenser prevents the loss of very volatile solvents so long as their boiling points are greater than $-78\ °C$, which is the sublimation point of solid carbon dioxide. Do not allow the ethanol to spill into the reaction flask. During the course of the reaction, an insulating coating of ice should be allowed to accumulate on the outside of the reaction flask. Monitor the amount of dry ice in the condenser and add more when necessary.

Weigh out 2.3 g (0.1 mol) of sodium in the following way. Remove a piece of sodium from the oil in which it is stored, and use a paper towel to soak up most of the oil that remains on the sodium lump. Do not use sodium that has developed a thick white coating of NaOH; instead, choose a piece that appears gray. On a paper towel, cut the sodium into small pieces and weigh the sodium by adding the pieces to a tared beaker containing heptane or similar nonvolatile alkane. It is important that you accurately weigh the Na, but it is not important that this amount be exactly 2.3 g. Adjust the amounts of the other reagents in accordance with the amount of Na used. Remove one of the stoppers on the three-neck flask and add 2.3 g (0.1 mol) of the freshly cut sodium metal. Replace the stopper. After approximately 10 min, the sodium dissolves completely to give a deep blue solution, which should be used immediately in the next step.

Dispose of any unwanted sodium scraps by adding them to a small beaker of anhydrous isopropanol, which reacts with sodium to form H_2 and $NaOC_3H_7$. After the sodium has completely reacted, the alcohol solution may be washed down the drain with a flush of water. Do *not* discard sodium metal in a waste basket or sink.

1,2-Bis(diphenylphosphino)ethane, $(C_6H_5)_2PCH_2CH_2P(C_6H_5)_2$ (dppe)

Add 13.1 g (0.05 mol) of solid triphenylphosphine to the Na-NH_3 solution. The addition should be done in five portions over the course of a few minutes to minimize frothing of the reaction. The solution changes from blue to the red-orange color characteristic of $NaPPh_2$. By gripping the necks of the flask, *gently* swirl the solution to dissolve any sodium that has deposited on the upper walls of the flask. After 30 min, cautiously add 4.9 g (0.05 mol) of dry NH_4Br to the reaction mixture with stirring. Prepare a solution of 2.47 g (0.025 mol) of 1,2-dichloro*ethane* in 20 mL of anhydrous ether. Add this solution to the orange $NaPPh_2$ solution and allow the mixture to stir for at least 10 min. Finally, remove the dry ice condenser, and allow the resultant pale orange mixture to evaporate. *Here is a good stopping point for the first day's work*: Liquid NH_3 does not evaporate quickly—simply leave the flask unstoppered in the hood until the next laboratory period.

To work up the reaction, add 100 mL of water and 75 mL of dichloromethane to the flask; stir the mixture well to dissolve most of the solid. Pour the two-phase mixture into a separatory funnel. Rinse the reaction flask with some additional dichloromethane and add this washing to the separatory funnel; drain the organic (lower) phase into a 250-mL round-bottom flask and dilute it with

100 mL of 95% ethanol. Concentrate this solution in a rotary evaporator to a total volume of about 75 mL. Filter off the colorless microcrystals (see Figure 13.1), wash them with a little ethanol, and allow them to air dry. Recrystallize the product by dissolving it in dichloromethane and, if necessary, filtering the solution to remove any solids. Dilute the dichloromethane solution with an equal volume of 95% ethanol and concentrate the mixture to about one-third its volume in a rotary evaporator. Collect the resulting crystals by filtration. Record the melting point.

Dichloro[1,2-bis(diphenylphosphino)ethane]nickel(II), NiCl$_2$(dppe)

Prepare a solution of 0.320 g (1.34 mmol) of NiCl$_2 \cdot 6$ H$_2$O in 50 mL of 95% ethanol. To this green solution add 0.54 g (1.34 mmol) of dppe. After the mixture has stirred for a few minutes, the product is fully formed and can be collected by filtration. Wash the orange solid with 20 mL of diethyl ether to remove unreacted dppe. Record the infrared (IR) spectrum of the product as a Nujol mull (see Experiment 19) and the proton nuclear magnetic resonance (^{1}H NMR) spectrum in CHCl$_3$ or CDCl$_3$ (consult your instructor). Integrate the NMR peaks due to the CH$_2$ and C$_6$H$_5$ groups (see Experiment 7).

REPORT

Include the following:
1. Yields of dppe and NiCl$_2$(dppe).
2. Integrated ^{1}H NMR spectra of dppe and NiCl$_2$(dppe) and assignments of peaks.
3. Infrared spectra of dppe and NiCl$_2$(dppe) with assignments of C–H, C–P, and C–C stretching bands and C–H bending bands.

PROBLEMS

1. What is the blue or bronze color of sodium in liquid NH$_3$ due to?
2. Suggest how the synthetic method described in this experiment could be modified to prepare a molecule with three phosphorus atoms.
3. Why is the glass-covered stir bar required for this procedure? Write a balanced equation describing what happens when the more common Teflon-coated stir bars come in contact with a sodium–ammonia solution.
4. Why do the PCH$_2$CH$_2$P protons appear as a triplet in the ^{1}H NMR spectrum of dppe?
5. Explain why it is important not to let ethanol leak into the reaction flask containing the sodium–ammonia solution.
6. Addition of a small amount of FeCl$_3$ to a solution of sodium in ammonia results in a color change from blue to colorless. Evaporation of the solution leaves a white solid. Explain these observations.

INDEPENDENT STUDIES

A. Prepare $P(C_6H_5)_2H$ by treatment of $NaP(C_6H_5)_2$ with excess NH_4Br. (Bianco, V. D.; Doronzo, S. *Inorg. Synth.* **1976**, *16*, 161.)

B. Prepare 2-diphenylphosphinobenzoic acid by the reaction of $NaP(C_6H_5)_2$ with 2-chlorobenzoic acid. (Hoots, J. E.; Rauchfuss, T. B.; Wrobleski, D. A. *Inorg. Synth.* **1982**, *21*, 175.)

C. Prepare $Mo(dppe)(CO)_4$ by the reaction of $Mo(CO)_6$ with dppe and $NaBH_4$ in refluxing ethanol. (Chatt, J.; Leigh, G. J.; Thankarajan, N. *J. Organomet. Chem.* **1971**, *29*, 105.)

D. Use dppe to prepare the dihydrogen complex $Fe(H_2)H(dppe)_2{}^+$. (Bautista, M. T.; Bynum, L. D.; Schauer, C. K. *J. Chem. Educ.* **1996**, *73*, 988.)

REFERENCES

Anhydrous Ammonia as a Solvent

Campbell, J.; Dixon, D. A.; Mercier, H. P. A.; Schrobilgen, G. J. *Inorg. Chem.* **1995**, *34*, 5798. Studies of the Zintl anion $Pb_9{}^{2-}$ in anhydrous ammonia.

Drake, G. W.; Kolis, J. W. *Coord. Chem. Rev.* **1994**, *137*, 131. Synthesis of Zintl compounds using liquid ammonia and related solvents.

Wagner, M. J.; Dye, J. L. *Ann. Rev. Mater. Sci.* **1993**, *23*, 223. Preparation of electrides and salts of negatively charged alkali metals.

Synthesis of Organophosphorus Ligands

Corbridge, D. E. C. *Phosphorus: An Outline of Its Chemistry, Biochemistry, and Technology*, Elsevier: Amsterdam, 1995. A broad overview of phosphorus chemistry.

Engel, R., Ed. *Synthesis of Carbon–Phosphorus Bonds*, CRC Press: Boca Raton, FL, 1988.

Hewertson, W.; Watson, H. R. *J. Chem. Soc.* **1962**, 1490. Early report on the synthesis of chelating diphosphine ligands.

Kosolapoff, G.; Maier, L. *Organic Phosphorus Chemistry*, Wiley: New York, 1972, Vols. I–VII. Useful survey of organophosphorus compounds with extensive tables of properties.

Newman, A. R.; Hackworth, C. A. *J. Chem. Educ.* **1986**, *63*, 817. Preparation of dppe.

NMR Properties of Phosphines

Carty, A. J.; Harris, R. K. *J. Chem. Soc., Chem. Commun.* **1967**, 235. Analysis of the NMR signals for dppe and related diphosphines.

Quin, L. D.; Verkade, J. G., Eds., *Phosphorus-31 NMR Spectral Properties in Compound Characterization and Structural Analysis*, VCH: New York, 1994.

Complexes of dppe and Related Ligands

Hartley, F. R. *The Chemistry of Organophosphorus Compounds*, Wiley: New York, 1990. Vols. I–III.

Hashigushi, S.; Fujii, A.; Takehara, J.; Ikariya, T.; Noyori, R. *J. Am. Chem. Soc.* **1995**, *117*, 7562. Asymmetric synthesis using chiral diphosphine ligands.

McAuliffe, C. A. *Transition Metal Complexes of Phosphorus, Arsenic and Antimony Ligands*, Wiley: New York, 1973. An older but useful review of the coordination complexes of phosphine ligands.

McAuliffe, C. A.; Levason, W. *Phosphine, Arsine and Stibine Complexes of the Transition Elements*, Elsevier: New York, 1979. A companion volume to the above.

Noyori, R. *Asymmetric Catalysis in Organic Synthesis*, Wiley: New York, 1994.

Ojima, I. *Catalytic Asymmetric Synthesis*, VCH: New York, 1993.

Undheim, K.; Bennecke, T. *Adv. Heterocyclic Chem.* **1995**, *62*, 306. Use of Ni and Pd diphos-phine complexes for cross-coupling of aryl halides.

Vaska, L.; Catone, D. L. *J. Am. Chem. Soc.* **1966**, *88*, 5324. Synthesis of $Ir(dppe)_2^+$ and derivatives.

Electrolytic Synthesis of $K_2S_2O_8$

Note: This experiment requires one 3- or 4-hour laboratory period. The saturated KHSO$_4$ solution should be prepared a day or week in advance of the experiment.

The industrial preparation of chemicals on a large scale by electrolysis is widely practiced. Passage of electrical current through aqueous solutions of NaCl yields the commercially important chemicals NaOH, Cl_2, and H_2. Electrolysis of anhydrous molten NaCl yields metallic sodium and Cl_2. Electrolytic methods are also used to obtain magnesium metal $MgCl_2$ and to win aluminum from Al_2O_3 (Hall–Heroult process). Most objects can be electroplated with Ag, Au, Pt, Cr, Ni, or Cu to increase their durability and attractiveness. Aside from the commercial applications, electrolysis is sometimes the only practical method of synthesizing certain chemicals. Fluorine, F_2, is one of these chemicals. Because F_2 is one of the strongest known oxidizing agents, F^- cannot be oxidized by chemical means to F_2 except under unusual circumstances. Voltage sources, however, easily oxidize F^-, usually as the fused KHF_2 salt, to F_2. Electrolysis is frequently the synthetic method of choice when the separation of products is difficult. For example, solutions of $CrCl_2$ may be prepared by reducing, with metallic Zn, an aqueous solution of $CrCl_3$:

$$2\,Cr^{3+} + 6\,Cl^- + Zn \longrightarrow 2\,Cr^{2+} + 6\,Cl^- + Zn^{2+}$$

The solution of $CrCl_2$ is, however, contaminated with $ZnCl_2$. To avoid contamination, an aqueous solution of $CrCl_3$ could be electrolyzed to give a pure solution of $CrCl_2$:

$$2\,CrCl_3 \longrightarrow 2\,CrCl_2 + Cl_2$$

In this experiment, potassium peroxydisulfate, $K_2S_2O_8$, will be prepared by the electrolysis of an aqueous solution of H_2SO_4 and K_2SO_4. The $S_2O_8{}^{2-}$ ion has been shown by X-ray studies to have the structure,

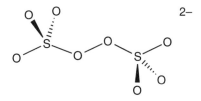

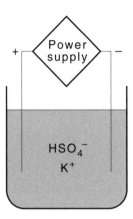

Figure 9-1
Schematic representation of an electrolysis cell.

in which the four oxygen atoms around each sulfur center are tetrahedrally arranged. In the solutions to be electrolyzed, K^+ and HSO_4^- are the major species present. An electrical current is passed through the solution, as shown schematically in Figure 9-1.

The reaction at the cathode is

$$2\,H^+ + 2\,e^- \longrightarrow H_2 \qquad E_{1/2} = 0.00 \text{ V versus SHE}$$

where SHE is the standard hydrogen electrode (see Experiment 5). The desired reaction at the anode is the oxidation:

$$2\,HSO_4^- \longrightarrow S_2O_8^{2-} + 2\,H^+ + 2\,e^- \qquad E_{1/2} = -2.05 \text{ V versus SHE}$$

It is obvious, however, that the oxidation of H_2O to O_2,

$$2\,H_2O \longrightarrow O_2 + 4\,H^+ + 4\,e^- \qquad E_{1/2} = -1.23 \text{ V versus SHE}$$

has a less negative oxidation potential and should occur in preference to that of HSO_4^- oxidation. The value of $E_{1/2} = -1.23$ V for the oxidation of water is not obtained from electrochemical measurements, but rather is a theoretical value obtained from the free energy change $\Delta G°$ for this process.

Fortunately, the oxidation of water at an electrode surface is very slow at this potential, and in practice much more than -1.23 V is required to liberate O_2 from water at a reasonable rate. The additional voltage is called an overvoltage. The rate of oxidation of water is greatly affected by the composition of the electrode at which the oxidation occurs. Thus, the overvoltage in 1 M KOH varies with the anode material as follows (current density = 1.0 amp cm^{-2}):

Anode	Overvoltage (V)
Ni	0.87
Cu	0.84
Ag	1.14
Pt	1.38

These overvoltages are not very reproducible and depend on the history of the anode material, but their differences suggest that the electrode participates directly in the oxidation process. Overvoltages are a familiar but poorly understood phenomenon. For the purposes of a synthetic chemist, the overvoltage permits the oxidation of some substances that could not be oxidized if the H_2O/O_2 couple exhibited no overvoltage. Owing to the high overvoltage observed for Pt, this material will be used as the anode material in the preparation of $K_2S_2O_8$.

To maximize the yield of $K_2S_2O_8$ and minimize the formation of O_2, it is advantageous to adjust other conditions of the electrolysis to increase the oxygen overvoltage. Relatively high currents will be used because overvoltages increase with current density (see below). Also, at low temperature, the rate of oxidation of H_2O will decrease, further increasing the overvoltage. Finally, high concentrations of HSO_4^- and low concentrations of H_2O will maximize $K_2S_2O_8$ yields. For these reasons, the electrolysis of HSO_4^- to form $S_2O_8^{2-}$ will be carried out using (1) a Pt electrode, (2) high current density, (3) low temperature, and (4) a saturated solution of HSO_4^-. Careful consideration of factors such as these has allowed the commercial electrolytic preparation and purification of chemicals on a large scale.

As in any electrolytic preparation, there is always the possibility that the product generated at the anode will diffuse to the cathode and be reduced to the starting material. Generally, the cathode and anode compartments must be separated and connected by a bridge to prevent this from occurring. In this experiment, the $S_2O_8^{2-}$ that is produced at the anode would be expected to diffuse toward the cathode, where it would be the most readily reduced species in solution; it would be immediately reduced to HSO_4^-. Fortunately, $K_2S_2O_8$ is quite insoluble in water, and it precipitates from solution before it reaches the cathode.

The Pt anode consists of a relatively small diameter wire (0.065 cm; 22 gauge). Knowing the diameter of the anode wire and the length of it coming in contact with the HSO_4^- solution, it is possible to calculate the current density, which is defined as

$$\text{Current density} = \text{current/area of anode}$$

The desired current density for this synthesis is 1.0 amp cm^{-2}. The number of amperes to be passed through the anode should be chosen to give this current density.

The amount of product generated will, of course, depend on the total number of electrons passed through the solution. Because the product of the current I (in amps) and the time of electrolysis t (in seconds) gives coulombs (C) of electricity, and 96,500 C oxidize (or reduce) one equivalent of reactant, the theoretical yield of product will be

$$\text{Theoretical yield} = \frac{\text{coulombs passed}}{96,500 \text{ coulombs/equiv}} (\text{equiv wt}) = \frac{I\,t}{96,500}(\text{equiv wt})$$

The actual yield is frequently less than this number of grams because of side reactions, and so percentage yields are usually evaluated. In electrochemistry, percentage yield is called current efficiency:

$$\text{Percentage yield} = \text{current efficiency} = \frac{\text{actual yield}}{\text{theoretical yield}} \times 100$$

Salts of the peroxydisulfate ion, $S_2O_8{}^{2-}$, are relatively stable but in acidic solution they react to give hydrogen peroxide, H_2O_2:

$$O_3S\text{--}O\text{--}O\text{--}SO_3{}^{2-} + 2\,H^+ \longrightarrow HO_3S\text{--}O\text{--}O\text{--}SO_3H$$

$$HO_3S\text{--}O\text{--}O\text{--}SO_3H + H_2O \longrightarrow HO_3S\text{--}O\text{--}OH + H_2SO_4$$

$$HO_3S\text{--}O\text{--}OH + H_2O \longrightarrow H_2SO_4 + H_2O_2$$

Under certain conditions, it is possible to stop the reaction at the intermediate peroxymonosulfuric acid, HO_3SOOH, but in the commercial preparation of H_2O_2 the reactions are driven to completion by distilling off the hydrogen peroxide.

The $S_2O_8{}^{2-}$ ion is one of the strongest known oxidizing agents and is even stronger than H_2O_2:

$$S_2O_8{}^{2-} + 2\,H^+ + 2\,e^- \longrightarrow 2\,HSO_4{}^- \qquad E_{1/2} = 2.05 \text{ V}$$

$$H_2O_2 + 2\,H^+ + 2\,e^- \longrightarrow 2\,H_2O \qquad E_{1/2} = 1.77 \text{ V}$$

It will oxidize many elements to their highest oxidation states, as will be illustrated by the reactions that you will carry out with $K_2S_2O_8$. For example, Cr^{3+} may be oxidized to $Cr_2O_7{}^{2-}$ according to the equation:

$$3\,S_2O_8{}^{2-} + 2\,Cr^{3+} + 7\,H_2O \longrightarrow 6\,SO_4{}^{2-} + Cr_2O_7{}^{2-} + 14\,H^+$$

Like most $S_2O_8{}^{2-}$ oxidations, the reaction is relatively slow, but is catalyzed by Ag^+ ions. A kinetic study of the Ag^+ catalyzed reaction suggests that the initial step is the oxidation of Ag^+ to the very reactive Ag^{3+} ion:

$$S_2O_8{}^{2-} + Ag^+ \longrightarrow 2\,SO_4{}^{2-} + Ag^{3+}$$

This Ag^{3+} ion then rapidly reacts with Cr^{3+} to give $Cr_2O_7{}^{2-}$ and regenerate Ag^+:

$$2\,Cr^{3+} + 3\,Ag^{3+} + 7\,H_2O \longrightarrow Cr_2O_7^{2-} + 3\,Ag^+ + 14\,H^+$$

The details of these reactions are unknown, but $S_2O_8^{2-}$ oxidations frequently proceed by first breaking the O–O bond to give the highly reactive radical SO_4^-, which carries out the further oxidations.

The strong oxidizing powers of $S_2O_8^{2-}$ have also permitted the syntheses of coordination complexes of silver in the unusual oxidation state of +2. The synthesis to be carried out in this Experiment is that of the complex $[Ag(py)_4]S_2O_8$, where py = pyridine:

$$2\,Ag^+ + 3\,S_2O_8^{2-} + 8\,py \longrightarrow 2\,[Ag(py)_4]S_2O_8 + 2\,SO_4^{2-}$$

The cation $Ag(py)_4^{2+}$ has a square planar geometry, analogous to that of $Cu(py)_4^{2+}$. In this reaction, the $S_2O_8^{2-}$ serves both as an oxidant and as a counteranion to precipitate the $Ag(py)_4^{2+}$ ion.

EXPERIMENTAL PROCEDURE

Electrolysis Cell

Cell construction is very simple. The anode is made by sealing a 0.065-cm diameter (22 gauge) platinum wire into 6-mm glass tubing. The length of anode that is in contact with the solution is about 6 cm. The cathode is a Pt wire wound around glass tubing. Insert the electrode assembly into a cork or rubber stopper that is loosely fitted into the approximately 2×20-cm test tube. The stopper should have a second small hole to allow gaseous reaction products to escape from the system. An adjustable power supply conveniently provides the 1.0 amp cm^{-2} current density required for the $K_2S_2O_8$ preparation. Note that this amperage level is dangerous, and all electrode connections should be made with care.

Potassium Peroxydisulfate, $K_2S_2O_8$

Safety note: hot sulfuric acid can cause severe burns. Do not touch the electrolysis apparatus while current is flowing.

The following preparation of a saturated solution of $KHSO_4$ should be done days in advance of the electrolysis. Carefully add 60 mL of concentrated H_2SO_4 to 150 mL of well-stirred water (*Caution*: Heat is evolved). Heat this solution to about 50 °C on a hot plate. Prepare a saturated solution of $KHSO_4$ by adding K_2SO_4 to the hot H_2SO_4 solution until no more dissolves. Then cool the solution to 0 °C in an ice bath for 1–2 h to ensure that precipitation of excess K_2SO_4 is complete. Pour the cold solution into a stoppered bottle for storage.

On the day of the experiment, pour the supernatant solution into the electrolysis cell and immerse the cell in an ice bath (Fig. 9-2). Turn on the power supply (record the time and temperature) and adjust the amperage until the anode current density is 1 amp cm^{-2}. The amperage required will be determined

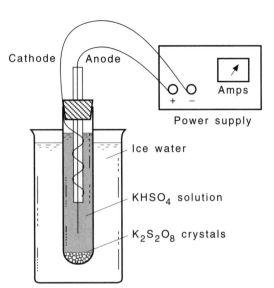

Figure 9-2
Electrolysis cell immersed in an ice bath.

by the surface area of the anode, as discussed earlier. Allow the current to flow for 30–45 min, during which time white crystals of $K_2S_2O_8$ collect on the bottom of the tube. The reaction will slow considerably toward the end of this period owing to depletion of HSO_4^-. The resistance of the solution to the current will generate sufficient heat to require replenishing the ice in the bath during electrolysis.

After the reaction period, turn off the power supply and record the time. Suction filter (Fig. 13-1) the $K_2S_2O_8$ crystals and wash them on the frit, first with 20 mL of 95% ethanol and finally with 20 mL of diethyl ether. Determine the yield. From the amperage and time, calculate the current efficiency. About 3 g of $K_2S_2O_8$ is required for subsequent reactions. If 3 g of $K_2S_2O_8$ is not obtained, add fresh HSO_4^- solution to the cell and repeat the synthesis.

Reactions of $K_2S_2O_8$
Prepare a saturated solution of $K_2S_2O_8$ by dissolving about 0.75 g of $K_2S_2O_8$ in a minimum of water. Add some of this $K_2S_2O_8$ solution to each of the solutions listed below. What happens in each case? (Use test tubes as reaction vessels.)

1.   Reaction with an acidified KI solution. Warm slightly.
2.   Reaction with $MnSO_4 \cdot H_2O$ in dilute H_2SO_4 to which 1 drop of $AgNO_3$ solution is added. Warm gently.
3.   Reaction with $Cr_2(SO_4)_3 \cdot xH_2O$ in dilute H_2SO_4 to which 1 drop of $AgNO_3$ solution is added. Warm gently.
4.   Reaction with $AgNO_3$ solution.

For comparison, instead of adding $K_2S_2O_8$, treat each of the above solutions with 30% H_2O_2. What happens in each case?

Tetrapyridinesilver(II) Peroxydisulfate, $[Ag(py)_4]S_2O_8$

Add 1.4 mL (1.3 g, 16.4 mmol) of pyridine to a solution of 0.16 g (0.95 mmol) of $AgNO_3$ in 3.2 mL of water. With stirring, pour this solution into a solution of 2.0 g of $K_2S_2O_8$ in 135 mL of water. Let the solution stand for 30 min, and then collect the yellow product by suction filtration (Fig. 13-1). Wash the product with 10 mL of water and dry it in a desiccator. Calculate the percentage yield and record the infrared (IR) spectrum of a Nujol mull of $[Ag(py)_4]S_2O_8$ (see Experiment 19 for the preparation of a Nujol mull).

REPORT

Include the following:
1. Calculation of the amperage to be used to attain a current density of 1 amp cm^{-2}.
2. Current efficiency in the $K_2S_2O_8$ preparation.
3. Balanced equations for the reactions of $K_2S_2O_8$ and H_2O_2 with KI, $MnSO_4$, $Cr_2(SO_4)_3$, and $AgNO_3$.
4. Percentage yield of $[Ag(py)_4]S_2O_8$ and assignments of infrared absorption bands to vibrations in the compound.

PROBLEMS

1. Give at least two reasons for the low current efficiency in the preparation of $K_2S_2O_8$. How would you experimentally determine which of these factors was most important in reducing the current efficiency?
2. Draw a Lewis diagram (dot formula) for the $S_2O_8^{2-}$ ion.
3. The $Ag(py)_2^+$ complex is colorless, whereas $Ag(py)_4^{2+}$ is colored. Explain why this is expected.
4. From the standard oxidation potentials of $S_2O_8^{2-}$ and H_2O, would you expect $S_2O_8^{2-}$ to oxidize H_2O to O_2 and H^+? In fact, does this reaction occur? Why or why not?
5. Give the formula of a compound, besides $[Ag(py)_4]S_2O_8$, that contains silver in the +2 oxidation state.
6. The $K_2S_2O_8$ that is produced may be contaminated with K_2SO_4 or $KHSO_4$. Suggest a method for determining the purity of the $K_2S_2O_8$ product.
7. Write an equation for the overall reaction that occurs during the electrolysis of the aqueous $KHSO_4$ solution.
8. Why can K_2SO_4 not be used in place of $KHSO_4$ in the preparation of $K_2S_2O_8$?
9. Why is it important that the cathode and anode not be too close to each other in the electrolysis solution?
10. If, instead of Pt, a Cu wire were used as the anode, would $K_2S_2O_8$ still be formed? Explain.

INDEPENDENT STUDIES

A. Analyze your $K_2S_2O_8$ product to determine its purity.

B. Measure the IR spectrum of $K_2S_2O_8$ and assign as many absorptions as possible.

C. Prepare the 2,2'-bipyridine (bipy) complex $[Ag(bipy)_2](NO_3)_2$ by electrolysis of a solution of $AgNO_3$ and bipy. (Thorpe, W. G.; Kochi, J. K. *J. Inorg. Nucl. Chem.* **1971**, *33*, 3958.)

D. Prepare AgO by electrolysis of $AgNO_3$. (Jolly, W. L. *The Synthesis and Characterization of Inorganic Compounds,* Prentice-Hall: Englewood Cliffs, NJ, 1970, p 448.)

E. Prepare amalgams of Ba or the lanthanides by electrolysis of metal salt solutions, using a Hg cathode. (Marklein, B. C.; West, D. H.; Audrieth, L. F. *Inorg. Synth.* **1939**, *1*, 11. Jukkola, E. E.; Audrieth, L. F.; Hopkins, B. S. *Inorg. Synth.* **1939**, *1*, 15.)

F. Prepare $KClO_3$ by electrolysis of KCl. (Pass, G.; Sutcliffe, H. *Practical Inorganic Chemistry*, 2nd ed., Halsted: New York, 1974, p 84.)

G. Prepare $[P(C_6H_5)_3H]_2CoCl_4$ by electrooxidation of cobalt. (Oldham, C.; Tuck, D. G. *J. Chem. Educ.* **1982**, *59*, 420.)

REFERENCES

$K_2S_2O_8$

Adams, D. M.; Raynor, J. B. *Advanced Practical Inorganic Chemistry*, Wiley: New York, 1965, p 122.

Chen, M. H.; Lee, S.; Liu, S.; Yeh, A. *Inorg. Chem.* **1996**, *35*, 2627. Kinetic studies of oxidations by peroxydisulfate.

Jolly, W. L. *The Synthesis and Characterization of Inorganic Compounds,* Prentice-Hall: Englewood Cliffs, NJ, 1970; pp 447–448.

Electrolytic Synthesis

Braam, J. M.; Carlson, C. D.; Stephens, D. A.; Rehan, A. E.; Compton, S. J.; Williams, J. M. *Inorg. Synth.* **1986**, *24*, 130. Experimental procedures for growing crystals on an electrode.

Chang, J.; Large, R. F.; Popp, G. in *Physical Methods of Chemistry*, Vol. 1, Part IIB, Weissberger, A.; Rossiter, B., Eds., Wiley: New York, 1971. Brief survey of electrochemical syntheses and experimental techniques.

Feltham, A. M.; Spiro, M. *Chem. Rev.* **1971**, *71*, 177. Factors that affect the behavior of platinized platinum electrodes.

Gewirth, A. A.; Niece, B. K. *Chem. Rev.* **1997**, *97*, 1129. In situ characterization of electrode surfaces.

Headridge, J. B. *Electrochemical Techniques for Inorganic Chemists,* Academic: New York, 1969. A brief, general discussion of electrolysis in inorganic synthesis.

Redox Reactions

Basolo, F.; Pearson, R. G. *Mechanisms of Inorganic Reactions*, 2nd ed., Wiley: New York, 1967, Chapter 6. Oxidation–reduction reactions of coordination compounds.

Espenson, J. H. *Chemical Kinetics and Reaction Mechanisms*, 2nd ed., McGraw-Hill: New York, 1995.

Jordan, R. B. *Reaction Mechanisms in Inorganic and Organometallic Systems*, 2nd ed., Oxford: New York, 1998.

Taube, H. *Science* **1984**, *226*, 1028. Mechanisms of oxidation–reduction reactions, primarily focusing on transition metal complexes.

Part III

COORDINATION CHEMISTRY

Ion Exchange Separation of Chromium Complexes

Note: This experiment requires one 3- or 4-hour laboratory period.

The separation of ions by ion exchange chromatography has touched almost every area of chemistry. In inorganic chemistry, ion exchange resins are widely employed for "softening" water, the process by which divalent cations such as Ca^{2+} and Mg^{2+} are replaced by Na^+. Also of note is the separation of the +3 ions of the different lanthanide metals by ion exchange chromatography. Because the lanthanide elements have very similar chemical properties, they are usually found mixed together in nature. Before the advent of ion exchange chromatography, salts of the lanthanide elements could only be separated from one another by tedious fractional crystallization techniques, often requiring hundreds of steps and weeks of work. In contrast, ion exchange chromatography allows the separation to be done in one step in a few hours.

The affinity of a particular ion for an ion exchange resin is governed by many factors, but the magnitude of the charge on the ion is one of the most important. On this basis, it is expected that the affinities of the following Cr^{III} cations for anionic sites in a resin will increase in the order:

$$CrCl_2(OH_2)_4^+ \; < \; CrCl(OH_2)_5^{2+} \; < \; Cr(OH_2)_6^{3+}$$

It is these differing affinities for a resin that allow the separation of these cationic complexes in this experiment.

Most commercially available ion exchange resins are insoluble polymers prepared by the copolymerization of styrene and various isomers of divinylbenzene. The divinylbenzene serves to crosslink the polystyrene chains. The relative amount of divinylbenzene that is used in the polymerization greatly affects the solubility and physical properties as well as the ion exchange properties of the eventual resin. In general, although not always, increased cross-linking reduces the sizes of the pores into which the ions must diffuse in order to reach ionic sites in the resin. Resins prepared by copolymerizing styrene/divinylbenzene mixtures rich in the latter monomer are accordingly more selective toward small ions. The resin to be used in this experiment (Dowex 50W-X8) is made by copolymerizing a

103

mixture of 92% styrene and a moderate amount, 8%, of divinylbenzene; the latter number is indicated by the "X" number of the resin. The resin is then sulfonated by treating it with sulfuric acid. This reaction places sulfonic acid groups, $-SO_3H$, on some of the benzene rings:

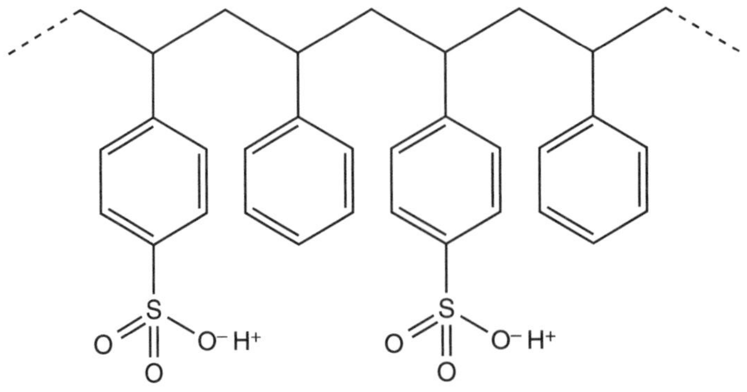

Figure 10-1
Structure of sulfonated polystyrene cation exchange resin.

Cation exchange resins that contain $-CO_2H$ or $-PO_3H_2$ groups are also available. Anion exchange resins usually contain the trimethylammonium group, $-N(CH_3)_3{}^+$.

The resin used in this experiment is supplied in the hydrogen ion (H^+) form, which is represented as $RSO_3{}^-H^+$. Other cations, M^+, that may be added to the resin will also have an affinity for the sulfonate groups and will displace the H^+ ions to a greater or lesser extent depending on the nature of the cation. In general, the equilibrium

$$RSO_3{}^-H^+ + M^+ \rightleftharpoons RSO_3{}^-M^+ + H^+$$

will be established. The value of the equilibrium constant for this reaction will be unique for any given M^+, and the position of the equilibrium will depend on the relative concentrations of M^+ and H^+ in solution. When the solution has a low H^+ concentration, M^+ will associate with the $RSO_3{}^-$ groups to the greatest extent. By increasing the H^+ concentration, it is possible to displace the M^+ ion from the resin. For two different cations, $M_1{}^+$ and $M_2{}^+$, displacement of the cation that has the lower affinity for the resin ($M_1{}^+$) will occur at relatively low H^+ concentrations, whereas higher acid concentrations will be required to displace the more tightly bound cation ($M_2{}^+$).

As previously mentioned, we expect $CrCl_2(OH_2)_4{}^+$ to be less strongly associated with the resin than $CrCl(OH_2)_5{}^{2+}$ or $Cr(OH_2)_6{}^{3+}$. The separation of the three ions in this experiment is based on the greater affinity of the more highly charged cations for the resin and on the higher H^+ concentrations that are required for their removal. If a solution of low acidity (2×10^{-3} M $HClO_4$) containing $CrCl_2(OH_2)_4{}^+$, $CrCl(OH_2)_5{}^{2+}$, and $Cr(OH_2)_6{}^{3+}$ is placed on a column of $RSO_3{}^-H^+$ resin (Fig. 10-1), all of the complex ions will strongly adhere. None of them may be eluted from the resin by passing $2 \times 10^{-3}\ M\ HClO_4$ through the column. On increasing the H^+ concentration by using $0.1\ M\ HClO_4$, the equilibrium will be sufficiently displaced to liberate the most weakly held cation, $CrCl_2(OH_2)_4{}^+$, which will be eluted from the column. It is necessary to increase the H^+ concentration to $1.0\ M\ HClO_4$ to force $CrCl(OH_2)_5{}^{2+}$ to migrate down the column. Finally, $4.0\ M\ HClO_4$ is required to displace the very strongly bound $Cr(OH_2)_6{}^{3+}$ from the resin.

The Cr^{III} ions that will be separated are derived from the compound that is usually designated as $CrCl_3 \cdot 6\ H_2O$. An X-ray structural investigation of it shows that this complex is actually $trans$-$[CrCl_2(OH_2)_4]Cl \cdot 2\ H_2O$. In the solid state, it contains discrete $trans$-$CrCl_2(OH_2)_4{}^+$, Cl^-, and H_2O units. The uncoordinated molecules of H_2O are hydrogen bonded to each other and to the H_2O molecules in the complex. In aqueous solution, $trans$-$CrCl_2(OH_2)_4{}^+$ undergoes aquation to form $CrCl(OH_2)_5{}^{2+}$.

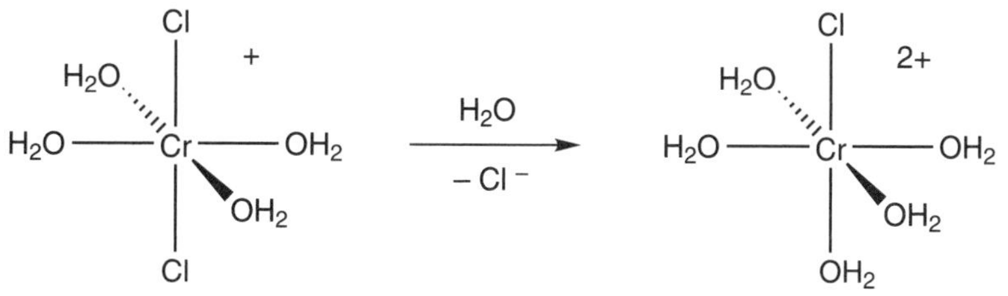

In $1.0\ M$ HCl solution at $25\,°C$, this reaction proceeds with a half-life of approximately 2.5 h. The product undergoes further aquation under the same conditions with a half-life of approximately 700 h to give $Cr(OH_2)_6{}^{3+}$. At lower acid concentrations, the reaction occurs considerably faster:

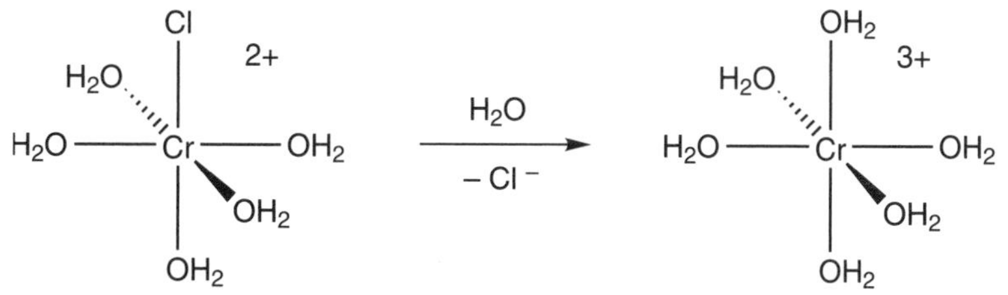

Because of this last reaction, the preparation of pure $CrCl(OH_2)_5{}^{2+}$ is quite difficult. The difference in affinity of these three complexes for a cation exchange resin, however, has permitted the isolation of pure solutions of $CrCl(OH_2)_5{}^{2+}$.

In this experiment, a solution of $trans$-$CrCl_2(OH_2)_4{}^+$ will be converted to $CrCl(OH_2)_5{}^{2+}$, and then to $Cr(OH_2)_6{}^{3+}$. You will use ion exchange chromatography to obtain pure solutions of each complex, and you will characterize the complexes by their visible spectra. From a knowledge of the spectra of these ions, you will then attempt to isolate and identify the ions that are present in an aged aqueous solution of $trans$-$CrCl_2(OH_2)_4{}^+$.

EXPERIMENTAL PROCEDURE

Safety note: Perchloric acid, $HClO_4$, is a strong oxidant and can react explosively with some organic compounds such as acetone. In the present experiment, which is conducted entirely in aqueous solutions, there should be no hazard as long as the solutions are kept away from organic solvents. At the end of the experiment, place all the solutions that contain chromium and perchloric acid into a labeled waste receptacle for them. The solutions should never be poured down the sink.

Preparation of Ion Exchange Column

A buret (about 10-mm inner diameter) may be conveniently used in this experiment (Fig. 10-2). Fill the buret three-quarters full with distilled water. Push a

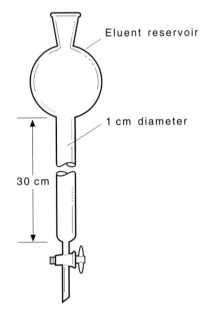

Figure 10-2
Apparatus for ion-exchange chromatography.

small plug of cotton or glass wool to the bottom of the buret with a rod. With the stopcock on the buret open and a large beaker positioned under the tip of the buret, pour into the buret a slurry of Dowex 50W-X8 (50–100 mesh, H^+ form) cation exchange resin in deionized water until a final resin height of approximately 15 cm is achieved. Allow water to pass through the resin until the effluent is colorless (this should take less than 100 mL). Then lower the water level so that it coincides with the top of the resin, and close the stopcock. If the water level drops below the top of the resin, channels will develop. Because channeling reduces the separation efficiency of a resin, never allow the resin to dry.

Prepare 200 mL each of 0.1, 1.0, and 4.0 M solutions of $HClO_4$ (see Appendix 1). These solutions will be used to elute the desired complexes.

trans-Dichlorotetraaquochromium(III) Ion, *trans*-$CrCl_2(OH_2)_4^+$

Prepare a 0.35 M Cr^{III} solution by dissolving 2.33 g (8.7 mmol) of commercial $CrCl_3 \cdot 6H_2O$ in a mixture of 25 mL of H_2O and one drop of concentrated

$HClO_4$. *Do not* heat the solution to dissolve the solid. *Important*: Record the time of day (i.e., hour and minute) when the dissolution of the $CrCl_3 \cdot 6 H_2O$ is performed. Portions of this Cr^{III} solution will be used throughout the experiment.

Immediately after this $0.35\ M\ Cr^{III}$ solution is prepared, add 5 mL of it to the cation exchange column previously prepared, and drain until the solution and resin levels are the same. Elute with $0.1\ M\ HClO_4$. When the most intense portion of the broad green band begins to drip out of the buret, collect a 5-mL fraction and immediately record the spectrum of the fraction on a visible spectrophotometer in the wavelength range from 350 to 800 nm, using glass cells of 1-cm path length (see Experiment 13 for comments on UV–vis spectrophotometers).

Chloropentaaquochromium(III) Ion, $CrCl(OH_2)_5^{2+}$

When $CrCl_2(OH_2)_4^+$ is warmed briefly in aqueous solution, much of it is converted to the cation $CrCl(OH_2)_5^{2+}$. Place an Erlenmeyer flask containing 5 mL of the $0.35\ M\ Cr^{III}$ solution in a boiling hot water bath for 3 min. (Do not allow the solution in the flask itself to come to a boil.) After the 3 min are up, remove the Erlenmeyer flask from the bath and immediately add 5 mL of distilled water. Pour the entire solution into the same buret that was used above. Drain the solution to the resin level, and flush the column with $0.1\ M\ HClO_4$ until the unreacted $CrCl_2(OH_2)_4^+$ has been eluted. The desired complex, $CrCl(OH_2)_5^{2+}$, is then eluted with $1.0\ M\ HClO_4$. Collect 5 mL of the most intensely colored fraction that elutes. Measure the visible spectrum of the solution.

Hexaaquochromium(III) Ion, $Cr(OH_2)_6^{3+}$

Combine 5 mL of the $0.35\ M\ Cr^{III}$ solution with 5 mL of distilled water and boil for 5 min. Add all of this solution to the same buret and drain it until the solution level reaches that of the resin. First rinse the column with $1.0\ M\ HClO_4$ to remove any unreacted $CrCl_2(OH_2)_4^+$ or $CrCl(OH_2)_5^{2+}$. Then elute the complex, $Cr(OH_2)_6^{3+}$, with $4.0\ M\ HClO_4$. Collect 5 mL of the intensely colored portion of the eluted $Cr(OH_2)_6^{3+}$ solution and then record its visible spectrum.

Separation of $CrCl_2(OH_2)_4^+$, $CrCl(OH_2)_5^{2+}$, and $Cr(OH_2)_6^{3+}$ from a Mixture

While these ions were being purified and characterized by their visible spectra, the $CrCl_2(OH_2)_4^+$ in the original stock solution of $0.35\ M\ CrCl_2(OH_2)_4^+$ has been aquating. The amounts of $CrCl_2(OH_2)_4^+$, $CrCl(OH_2)_5^{2+}$, and $Cr(OH_2)_6^{3+}$ that are now present in solution depend on the length of time the solution has been standing, which should be a few hours. (If the solution is allowed to stand overnight, only $Cr(OH_2)_6^{3+}$ will remain.) The object of this portion of the experiment is to determine which complexes are now present in solution.

Prepare a new column of resin. Combine 5 mL of the $0.35\ M\ Cr^{III}$ solution with 5 mL of distilled water, and add the mixture to the buret. Record the time that has elapsed since the $0.35\ M\ Cr^{III}$ solution was prepared. First elute all of the $CrCl_2(OH_2)_4^+$ that might be present with $0.1\ M\ HClO_4$. Collect the entire fraction. Next elute all the $CrCl(OH_2)_5^{2+}$ that might be present with $1.0\ M$ $HClO_4$, and finally, elute all the $Cr(OH_2)_6^{3+}$ that might be present with $4.0\ M$ $HClO_4$. Measure the volumes of each of the fractions, record their visible spectra, and measure the absorbance at an appropriate wavelength.

By comparing the spectra of the fractions obtained in this separation with those determined for the known compounds in the previous section, note which complexes are now present in the solution and roughly which complexes are in greatest and least abundance. By assuming that the extinction coefficients of the three complexes are approximately the same, determine the mole fraction of each species present in the "aged" Cr^{III} solution.

REPORT

Include the following:
1. Visible spectra of $CrCl_2(OH_2)_4^+$, $CrCl(OH_2)_5^{2+}$, and $Cr(OH_2)_6^{3+}$.
2. Wavelengths of the absorption maxima for the three compounds. Comment on the relative ligand field strengths of Cl^- and H_2O.
3. Molar fractions of the species present in the "aged" 0.35 M Cr^{III} solution. Include the time of aging.

PROBLEMS

1. The aquation of $CrCl_2(OH_2)_4^+$ to $CrCl(OH_2)_5^{2+}$ is catalyzed by $Cr(OH_2)_6^{2+}$. Write a reasonable mechanism that will account for the catalytic role of this chromium(II) ion.
2. The conversion of $CrCl(OH_2)_5^{2+}$ to $Cr(OH_2)_6^{3+}$ in water is catalyzed by Hg^{2+}. Write a mechanism for this catalysis.
3. Qualitatively account for the observation that the uncatalyzed aquation of $CrCl_2(OH_2)_4^+$ to $CrCl(OH_2)_5^{2+}$ occurs much faster than that of $CrCl(OH_2)_5^{2+}$ to $Cr(OH_2)_6^{3+}$ under the same conditions.
4. A 10-mL solution of 0.5 M $CrCl(OH_2)_5^{2+}$ is allowed to aquate to $Cr(OH_2)_6^{3+}$. To determine an approximate rate of reaction, the amounts of $CrCl(OH_2)_5^{2+}$ and $Cr(OH_2)_6^{3+}$ present after a certain time are measured by pouring the solution onto a cation exchange resin in the H^+ form and then titrating the displaced H^+ with base. If 80 mL of 0.15 M NaOH is required to neutralize the liberated H^+, what were the concentrations of $CrCl(OH_2)_5^{2+}$ and $Cr(OH_2)_6^{3+}$ in the solution?
5. Explain how you would separate $K_4[Fe(CN)_6]$ from $K_3[Fe(CN)_6]$.
6. Why was $HClO_4$ rather than HCl used to elute the Cr^{III} complexes from the column?
7. Why was visible rather than IR spectroscopy used to characterize the complexes in this experiment?

INDEPENDENT STUDIES

A. Prepare, isolate, and characterize $[CrCl(OH_2)_5]Cl_2 \cdot H_2O$ and $[Cr(OH_2)_6]Cl_3$. (Barbier, J. P.; Kappenstein, C.; Hugel, R. *J. Chem. Educ.* **1972**, *49*, 204.)
B. Determine the rate of aquation of $CrCl(OH_2)_5^{2+}$ to give $Cr(OH_2)_6^{3+}$. (Swaddle, T. W.; King, E. L. *Inorg. Chem.* **1965**, *4*, 532.)
C. Prepare and separate by ion exchange chromatography the cis and trans isomers of $Co(IDA)_2^-$, where $IDA^{2-} = HN(CH_2CO_2^-)_2$. (Weyh, J. A. *J. Chem. Educ.* **1970**, *47*, 715.)

D. Separate the rare earth metals from a mixture by eluting with a citric acid/ citrate buffer. (Thompson, R., Ed., *Specialty Inorganic Chemicals*, Specialty Publication 40, Royal Society of Chemistry: London, 1981, pp 403–440; Swaddle, T. W. *Inorganic Chemistry: An Industrial and Environmental Perspective*, Academic: New York, 1997, pp. 366–367.)

REFERENCES

$CrCl_2(OH_2)_4^+$, $CrCl(OH_2)_5^{2+}$, and $Cr(OH_2)_6^{3+}$

Finholt, J. E.; Caulton, K. G.; Libbey, W. J. *Inorg. Chem.* **1964**, *3*, 1801. Synthesis and UV–vis spectra of $CrCl(H_2O)_5^{2+}$.

Johnson, H. B.; Reynolds, W. L. *Inorg. Chem.* **1963**, *2*, 468. Rate of aquation of $CrCl_2(H_2O)_4^+$.

Nyburg, S. C.; Šoptrajanov, B.; Stefov, V.; Petruševski, V. M. *Inorg. Chem.* **1997**, *36*, 2248. X-ray crystal structure of *trans*-$CrCl_2(H_2O)_4^+$.

Swaddle, T. W.; King, E. L. *Inorg. Chem.* **1965**, *4*, 532. Rate of aquation of $CrCl(H_2O)_5^{2+}$.

Magini, M. *J. Chem. Phys.* **1980**, *73*, 2499. Structural study of Cr^{III} in aqueous solution.

Beattie, J. K.; Best, S. P.; Skelton, B. W.; White, A. H. *J. Chem. Soc., Dalton Trans.* **1981**, 2105. X-ray crystal structure of $Cr(H_2O)_6^{3+}$.

Best, S. P.; Armstrong, R. S.; Beattie, J. K. *Inorg. Chem.* **1980**, *19*, 1958. Infrared spectrum of $Cr(H_2O)_6^{3+}$.

Ion Exchange Techniques

Dorfner, K. *Ion Exchangers: Properties and Applications*, 3rd ed., Ann Arbor Science: Ann Arbor, MI, 1972.

Druding, L. F.; Kauffman, G. B. *Coord. Chem. Rev.* **1968**, *3*, 409. Chromatography of coordination compounds.

Harland, C. E. *Ion Exchange*, 2nd ed., Royal Society of Chemistry: Cambridge, UK, 1994.

Strauss, S. D. *Power* **1993** (June), 17. A survey of water treatment technologies including the use of ion exchange resins.

Walton, H. F.; Rocklin, R. D. *Ion Exchange in Analytical Chemistry*, CRC Press: Boca Raton, FL, 1990.

General References to Coordination Compounds

Basolo, F.; Johnson, R. *Coordination Chemistry*, 2nd ed., Science Reviews, 1986. Elementary introduction (paperback).

Cotton, F. A.; Wilkinson, G. *Advanced Inorganic Chemistry*, 5th ed., Wiley: New York, 1988. Contains considerable descriptive chemistry.

Wilkinson, G.; McCleverty, J. A.; Gillard, R. D., Eds. *Comprehensive Coordination Chemistry*, Pergamon: New York, 1987. A seven-volume encyclopedia of coordination compounds.

Metal–Metal Quadruple Bonds

Note: This experiment requires 4 hours for Part A (split between two lab sessions) and 4 hours for Part B.

The concept of bond order is one of the most fundamental in chemistry. Single, double, and triple bonds have long been known and are well understood. In organic compounds, the maximum bond order of 3 is a direct consequence of the number of s and p orbitals and their spacial orientation. In compounds of the transition elements, however, the d orbitals can also participate in bonding. As a result, bond orders greater than 3 are possible in certain compounds of the transition elements. In 1964 there appeared the first report of a compound with a quadruple bond, that is, a bond of order 4. In this experiment, you will prepare and characterize an interesting family of molybdenum compounds that have quadruple bonds.

Metal–metal single bonds are often seen in transition metal clusters in which the metals are in low oxidation states (see Experiments 17 and 18). The presence of M–M bonding is supported by magnetic and structural evidence. For example, manganese forms a carbonyl compound of stoichiometry $Mn(CO)_5$; if monomeric, this 17-electron compound should be paramagnetic. Experimentally, however, this compound is diamagnetic owing to the formation of a metal–metal bond. Crystallographic studies show that the compound is a dimer, $[Mn(CO)_5]_2$, in which the two Mn atoms are connected by a single bond that is 2.9 Å long.

The bond order of a metal–metal bond depends on the ligands attached to the metals and also on the number of d electrons available. If there are few d electrons, then the M–M bond order will be small. If there are many d electrons, then M–M antibonding orbitals will be filled and the metal–metal bond order will also be small. The optimal situation for the formation of quadruple metal–metal bonds occurs when there are exactly four d electrons on each metal center.

One of the most famous compounds with a quadruple bond is tetra(acetato)dimolybdenum(II), $Mo_2(O_2CCH_3)_4$. In this diamagnetic molecule, each Mo^{II} atom has four valence d electrons. Together, the two molybdenum centers have a total of eight d electrons, all of which occupy orbitals that are metal–metal bonding. The molecular structure of $Mo_2(O_2CCH_3)_4$ is shown in

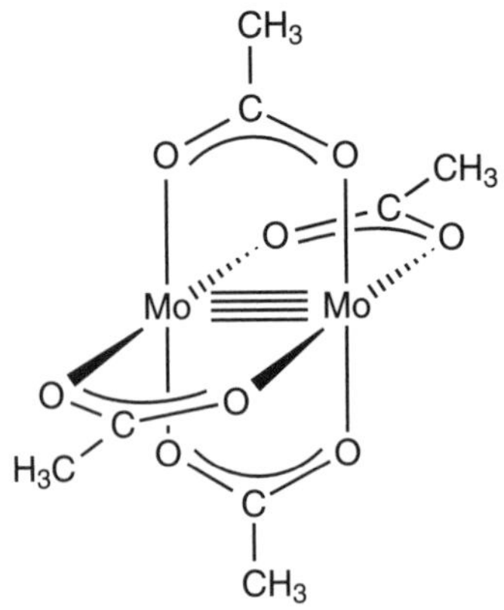

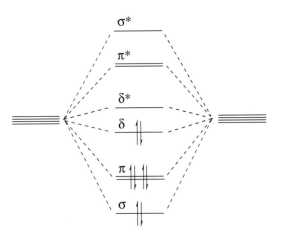

Figure 11-1
Structure of $Mo_2(O_2CCH_3)_4$.

Figure 11-1. The four acetate groups are just the right length to bridge between the two metal centers.

The molecular orbital (MO) diagram of compounds with quadruple metal–metal bonds is shown in Figure 11-2. The σ and π MOs (and their antibonding σ^*

Figure 11-2
Molecular orbital diagram of compounds with quadruple metal–metal bonds.

and π^* counterparts) are roughly similar to those found in unsaturated organic compounds, except that they are made by overlap of two d orbitals rather than two p orbitals. The new feature of this MO diagram is the presence of the δ (and δ^*) MOs. These are formed by overlap of two d_{xy} orbitals (one from each metal center). Quadruply bonded compounds have the electronic configuration $\sigma^2\pi^4\delta^2$.

In the first part of this experiment, you will synthesize tetra(acetato)-dimolybdenum(II) by refluxing the zero-valent complex $Mo(CO)_6$ in acetic acid.

Molybdenum hexacarbonyl is a convenient starting material because the carbon monoxide ligands are readily lost and the Mo^0 center is easily oxidized once these acceptor ligands have been removed. New compounds that contain quadruple metal–metal bonds can be prepared by replacing the acetate groups in $Mo_2(O_2CCH_3)_4$ with other ligands. In the second part of this experiment, you will treat $Mo_2(O_2CCH_3)_4$ with concentrated HCl to form the octachloro-dimolybdate(II) anion, $Mo_2Cl_8^{4-}$, which is isolated as its potassium salt. This red-purple compound is a rare example of an unbridged quadruple bond. Treatment of $K_4Mo_2Cl_8$ with tributylphosphine displaces four of the chloride groups and forms the intensely violet compound $Mo_2Cl_4[P(C_4H_9)_3]_4$. These reactions are summarized below:

$$2\,Mo(CO)_6 + 4\,HO_2CCH_3 \longrightarrow Mo_2(O_2CCH_3)_4 + 12\,CO + 2\,H_2$$

$$Mo_2(O_2CCH_3)_4 + 4\,HCl + 4\,KCl \longrightarrow K_4Mo_2Cl_8 + 4\,HO_2CCH_3$$

$$K_4Mo_2Cl_8 + 4\,P(C_4H_9)_3 \longrightarrow Mo_2Cl_4[P(C_4H_9)_3]_4 + 4\,KCl$$

The structures of $Mo_2Cl_8^{4-}$ and $Mo_2Cl_4[P(C_4H_9)_3]_4$ are shown below:

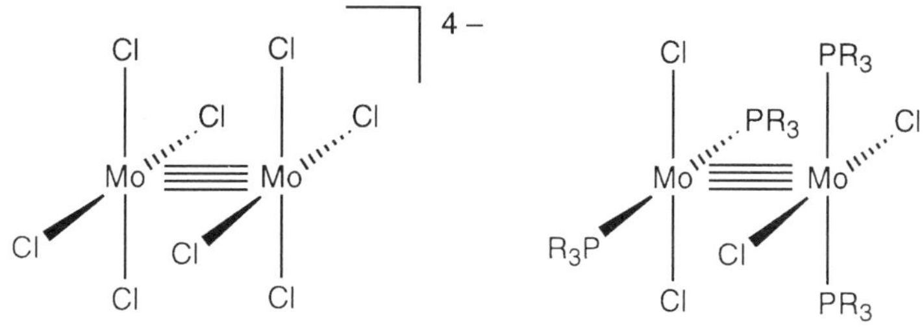

EXPERIMENTAL PROCEDURE

Part A

Tetra(acetato)dimolybdenum(II), $Mo_2(O_2CCH_3)_4$

Safety note: Acetic anhydride reacts violently with water and can spray acid into the eyes and onto clothes. Glassware that has been used to handle acetic anhydride should be rinsed in a hood or behind a shield. Carbon monoxide is evolved in this experiment. The reaction should be performed in a hood.

In a hood, equip a 250-mL two-neck round-bottom flask with a Teflon-coated stir bar and a water-cooled reflux condenser topped with a N_2 gas inlet. The reflux condenser should have a wide internal bore so that it will not plug up when the molybdenum carbonyl sublimes out of solution during the next step. If possible, use Apiezon H grease to lubricate the joints, because this grease is more resistant to the leaching action of refluxing acetic acid. To the N_2 flushed flask, add 100 mL of acetic acid, 7 mL of fresh acetic anhydride, and three boiling stones. Plug the second neck of the flask with a greased glass stopper. Firmly clamp all the joints together; failure to do this will cause the solvent to leak out and will ruin the experiment. Boil the solution under N_2 for 30 min with a rheostat-controlled heating mantle. Cool the solution to room temperature in an ice bath, remove the stopper, wipe the grease out of the joint, and then add 2.0 g (7.5 mmol) of molybdenum hexacarbonyl through the joint against a gentle N_2 counterflow. Replug the joint with a glass stopper lubricated with Apiezon H grease and clamp it firmly in place.

Begin refluxing the mixture. For the first hour or so, monitor the reaction; molybdenum hexacarbonyl tends to sublime into the reflux condenser. If this happens, turn off the water to the reflux condenser. The acetic acid should slowly begin to reflux higher up the condenser and wash down the $Mo(CO)_6$. When most of the white solid has been washed back into solution, *turn the water back on*. At this point, the reaction mixture can be left to reflux overnight; you can use any time remaining in the lab period to finish up other experiments. *Here is a good stopping point for the first day's work.* Make arrangements with your instructor to return to the laboratory tomorrow to turn the heating mantle off. The yield of product will be greatly reduced if the mixture is refluxed for more than 24 h.

After the mixture has refluxed for 24 h, turn off the heating mantle, and let the dark-colored solution cool to room temperature under N_2.

During your next lab session, filter the solution in air using a glass frit (see Figure 13-1). Wash the bright yellow needles with two 20-mL portions of 95% ethanol and two 20-mL portions of diethyl ether, and then dry the product in vacuum. Remove the three boiling stones, measure the yield, and characterize your product by infrared (IR) spectroscopy. The product oxidizes in air over a period of weeks, so for long-term storage the compound should be kept under N_2.

Part B

Potassium Octachlorodimolybdate(II), $K_4Mo_2Cl_8\cdot2H_2O$

In a hood, place a stirring bar into a 100-mL flask. Add 30 mL of concentrated HCl, 1.5 g (20.1 mmol) of potassium chloride, and 1.0 g (2.35 mmol) of $Mo_2(O_2CCH_3)_4$. Stir the solution until all of the yellow needles dissolve. Filter off the resulting red-purple precipitate on a glass frit, wash the precipitate twice with 20 mL of absolute ethanol, and if possible dry the product in vacuum. Characterize your product by IR spectroscopy.

Tetrachlorotetrakis(tributylphosphine)dimolybdenum(II), $Mo_2Cl_4[P(C_4H_9)_3]_4$

Equip a 100-mL two-neck flask with a magnetic stir bar and a reflux condenser. Use silicone grease to lubricate the joints. Then add 0.5 g (0.8 mmol) of $K_4Mo_2Cl_8\cdot2H_2O$, 20 mL of degassed methanol, and 1.0 mL (3.6 mmol) of tributylphosphine. Plug the open neck of the flask with a greased glass stopper.

Bring the mixture to reflux for 2 h with a rheostat-controlled heating mantle. During this time, the red $K_4Mo_2Cl_8$ should disappear and be replaced by a blue-purple microcrystalline precipitate. Cool the solution and collect the precipitate by filtration. Wash the precipitate twice with 20 mL of water and once with 10 mL of methanol, and if possible dry the precipitate in vacuum. Record the yield and characterize your product by IR and proton nuclear magnetic resonance (^{1}H NMR) spectroscopy (the latter in chloroform).

REPORT

Include the following:
1. Yields of $Mo_2(O_2CCH_3)_4$, $K_4Mo_2Cl_8$, and $Mo_2Cl_4[P(C_4H_9)_3]_4$.
2. Infrared spectra and interpretation.
3. ^{1}H NMR spectrum of $Mo_2Cl_4[P(C_4H_9)_3]_4$.

PROBLEMS

1. Which d orbitals are involved in the formation of the quadruple bonds in these compounds? What are the electronic transitions responsible for the intense colors of quadruply bonded compounds?
2. How do the Mo–Mo distances in these compounds compare with those in metallic molybdenum and with those in compounds that have single Mo–Mo bonds?
3. What is known about the strength of quadruple bonds compared with the strength of the triple bond in acetylenes and N_2?
4. Metal carboxylate compounds have two IR bands associated with the symmetric and antisymmetric O–C–O stretches. Where are these bands in the IR spectrum of $Mo_2(O_2CCH_3)_4$? Can you assign the other bands in the IR spectrum of this molecule?

INDEPENDENT STUDIES

A. Prepare the quadruply bonded dichromium complex $Cr_2(O_2CCH_3)_4 \cdot 2H_2O$ from $CrCl_3 \cdot 6H_2O$, zinc, and sodium acetate. (Ocone, L. R.; Block, B. P. *Inorg. Synth.* **1966**, *8*, 125; Jolly, W. L. *The Synthesis and Characterization of Inorganic Compounds*, Prentice-Hall: Englewood Cliffs, NJ, 1970, p 442.)
B. Prepare the quadruply bonded dirhenium compounds $Re_2Cl_8^{2-}$ or $Re_2(O_2CCH_3)_4Cl_2$. (Barder, T. J.; Walton, R. A. *Inorg. Synth.* **1985**, *23*, 116; Cotton, F. A.; Oldham, C.; Robinson, W. R. *Inorg. Chem.* **1966**, *5*, 1798.)
C. Record the Raman spectra of the quadruply bonded dimolybdenum complexes to determine the frequencies of the metal–metal stretching modes. (Ketteringham, A. P.; Oldham, C. *J. Chem. Soc., Dalton Trans.* **1973**, 1067; Ketteringham, A. P.; Oldham, C.; Peacock, C. J. *J. Chem. Soc., Dalton Trans.* **1976**, 1640; Hutchinson, B.; Morgan, J.; Cooper, C. B.; Mathey, Y.; Shriver, D. F. *Inorg. Chem.* **1979**, *18*, 2048.)

D. Record the UV–vis spectra of the quadruply bonded dimolybdenum compounds and identify the δ to δ^* transition. (Trogler, W. C.; Solomon, E. I.; Trajberg, I.; Ballhausen, C. J.; Gray, H. B. *Inorg. Chem.* **1977**, *16*, 828; Martin, D. S.; Newman, R. A.; Fanwick, P. E. *Inorg. Chem.* **1979**, *18*, 2511; Hopkins, M. D.; Gray, H. B. *J. Am. Chem. Soc.* **1984**, *106*, 2468.)

E. Prepare the trifluoroacetate complex $Mo_2(O_2CCF_3)_4$ by addition of trifluoroacetic acid to $Mo_2(O_2CCH_3)_4$. (Cotton, F. A.; Norman, J. G. *J. Coord. Chem.* **1971**, *1*, 161.)

REFERENCES

$Mo_2(O_2CCH_3)_4$, $K_4Mo_2Cl_8$, and $Mo_2Cl_4[P(C_4H_9)_3]_4$
Brencic, J. V.; Cotton, F. A. *Inorg. Chem.* **1970**, *9*, 351. Preparation of $K_4Mo_2Cl_8$.
San Filippo, J. Jr. *Inorg. Chem.* **1972**, *11*, 3140. Preparation of $Mo_2Cl_4[P(C_4H_9)_3]_4$.
Brignole, A. B.; Cotton, F. A. *Inorg. Synth.* **1972**, *13*, 81. Preparation of $Mo_2(O_2CCH_3)_4$.

Other References to Compounds with Metal–Metal Multiple Bonds
Cotton, F. A. *J. Chem. Educ.* **1983**, *60*, 713. An introduction to quadruply bonded compounds.
Cotton, F. A.; Walton, R. A. *Multiple Bonds between Metal Atoms*, 2nd ed., Clarendon: Oxford, UK, 1993. A thorough review of the field.
Gross, C. L.; Wilson, S. R.; Girolami, G. S. *Inorg. Chem.* **1995**, *34*, 2582. Synthesis of the $Os_2Br_8^{2-}$ anion.

Molecular Orbital Theory
Albright, T. A.; Burdett, J. K. *Problems in Molecular Orbital Theory*, Oxford: New York, 1992.
Albright, T. A.; Burdett, J. K.; Whangbo, M.-H. *Orbital Interactions in Chemistry*, Wiley: New York, 1985.
Cotton, F. A. *Chemical Applications of Group Theory*, 3rd ed., Wiley: New York, 1990.
DeKock, R. L.; Gray, H. B. *Chemical Structure and Bonding*, Benjamin/Cummings, Menlo Park CA, 1980.
Jean, Y.; Volatron, F.; Burdett, J. K. *An Introduction to Molecular Orbitals*, Oxford: New York, 1993.
Kettle, S. F. A. *Symmetry and Structure*, Wiley: New York, 1995.

The Paramagnetic Complex Mn(acac)$_3$

Note: This experiment requires one 3- or 4-hour laboratory period.

Substances that contain unpaired electrons are strongly attracted into a magnetic field and are said to be paramagnetic. In contrast, substances with no unpaired electrons are weakly repelled by such a field and are called diamagnetic. The repulsion seen for diamagnetic materials arises because electron pairs always generate a magnetic field that will oppose an applied field. Thus the magnetic properties of all compounds have a diamagnetic component, but the overall character will be paramagnetic if unpaired electrons are present because the latter interact so strongly with an external magnetic field.

The magnetism of most transition metal complexes is closely related to the number of unpaired electrons present on the metal center. For many complexes of first-row transition metals, the spin angular momentum of the unpaired electrons is almost entirely responsible for the attraction between the complex and the magnetic field. In other complexes, however, especially those of the second- and third-row transition metals, the attraction depends not only on the electron's spin but also on its orbital angular momentum. This additional contribution complicates the relationship between the attractive force and the number of unpaired electrons. In this experiment, you will measure the magnetic susceptibility of a coordination compound of a first-row transition metal in which the orbital angular momentum does not contribute significantly to the magnetic properties. From the result, you will calculate the number of unpaired electrons.

As we will see, the number of unpaired electrons in a transition metal complex is very useful information. First, the number of unpaired electrons is closely related to the oxidation state of the metal center. For example, copper compounds often have 0 or 1 unpaired electron, depending on whether the oxidation state of the copper atom is $+1$ or $+2$.

Second, the number of unpaired electrons may be helpful in assigning geometries to complexes. The geometries of four-coordinate Ni complexes, Ni^{II}L$_4$, have been extensively studied by this method. Complexes of this formula may have either square planar or tetrahedral geometries. According to crystal field theory, the five d orbitals on Ni split in different ways depending on the geometry of the complex (Fig. 12-1).

117

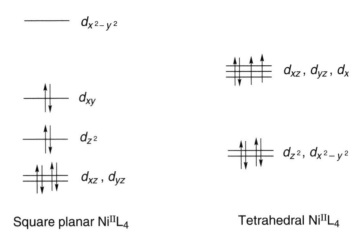

$d_{x^2-y^2}$

d_{xz}, d_{yz}, d_x

d_{xy}

d_{z^2}

$d_{z^2}, d_{x^2-y^2}$

d_{xz}, d_{yz}

Square planar $Ni^{II}L_4$ Tetrahedral $Ni^{II}L_4$

Figure 12-1
Orbital splitting diagrams for square planar and tetrahedral NiL_4 complexes.

The placement of the eight d electrons of Ni^{II} into the lowest energy orbitals leaves two unpaired electrons for tetrahedral complexes and no unpaired electrons for square planar complexes. Four-coordinate Ni complexes have been found by magnetic susceptibility studies to assume either the tetrahedral or the square planar configuration, depending on the ligands bound to the metal. There are even examples of Ni^{II} complexes that, in solution, exist as a mixture of tetrahedral and square planar forms.

Third, the number of unpaired electrons can provide information about metal-metal bonding. For example, the metal carbonyl complex $Fe_2(CO)_9$ has

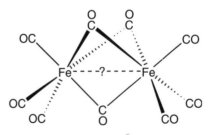

long been known from X-ray structural studies to have the following geometry The coordination geometry around each Fe atom is approximately octahedral. If the CO ligands are regarded as electrically neutral, then the Fe atom is in the zero oxidation state and will have eight valence electrons. Each Fe atom will bond with two electrons from each of the terminal CO groups and one electron from each of the bridging CO groups. Thus, each Fe atom is surrounded by a total of

17 bonding and valence electrons. Accordingly, one might expect that each Fe atom would be paramagnetic with one unpaired electron. Instead, the compound is diamagnetic, and a reasonable bonding proposal requires that there is an electron-pair, covalent bond between the Fe atoms (the dotted line in the drawing). The relatively short Fe–Fe distance in the compound ($2.52\,\text{Å}$) supports this suggestion.

Fourth, the number of unpaired electrons in a complex can also provide valuable information about the bonding between a metal and its ligands. For an octahedral transition metal complex, crystal field theory shows that the five d orbitals will split into a set of three orbitals at low energy (called the t_{2g} orbitals) and a set of two orbitals at higher energy (called the e_g orbitals). The energy gap between the t_{2g} and e_g orbitals is different for different ligands: Ligands such as carbon monoxide and cyanide give large energy gaps, whereas ligands such as fluoride and hydroxide give small energy gaps. A ranking of ligands based on the size of the energy gap they create is called the spectrochemical series; a shortened version of this series is shown below:

$$I^- \;<\; Br^- \;<\; Cl^- \;<\; F^- \;<\; OH^- \;<\; H_2O \;<\; NH_3 \;<\; NO_2^-$$

$$<\; PPh_3 \;<\; CH_3^- \;<\; CN^- \;<\; CO$$

Generally, ligands that are strong σ donors or strong π acceptors cause the largest splittings.

For octahedral complexes that possess between 4 and 7 d electrons, there are two different ways to distribute the electrons among the t_{2g} and e_g orbitals. Electrons tend to repel one another, and if the energy of the e_g orbitals is not too high, then the electrons will occupy as many different d orbitals as possible. In such cases, the electrons tend to arrange themselves with their spins parallel; this leads to what is known as a high-spin state. For a complex with five d electrons, the following splitting diagram is obtained:

e_g

t_{2g}

The magnetic properties of such a complex will show that it possesses five unpaired electrons. This situation is found for the aqua complex $Fe(H_2O)_6{}^{3+}$.

On the other hand, if the energy gap between the t_{2g} and e_g levels is large, then a more stable arrangement of the electrons is to place as many as possible into the low-energy t_{2g} orbitals. The Pauli exclusion principle requires that placing two electrons in the same d orbital forces them to have opposite spins; this leads to what is known as a low-spin state. For a complex with five d electrons, the following splitting diagram is obtained:

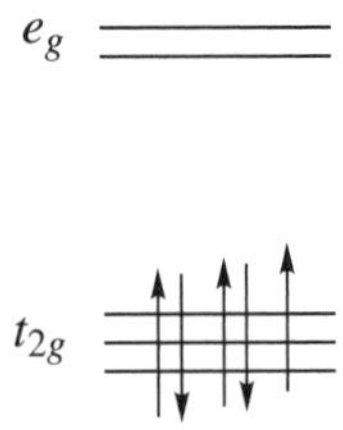

The magnetic properties of such a complex will show that it possesses only one unpaired electron. This electron configuration is found for the cyanide complex $Fe(CN)_6^{3-}$.

It is left as an excercise for the student to determine how the electrons arrange themselves in the high-spin and low-spin forms of complexes with 4, 6, or 7 d electrons.

In the present experiment, you will synthesize the six-coordinate manganese-(III) complex $Mn(acac)_3$, where acac stands for acetylacetonate (this ligand is also referred to as 2,4-pentanedionate and is sometimes abbreviated pd). The structure of the complex is shown below:

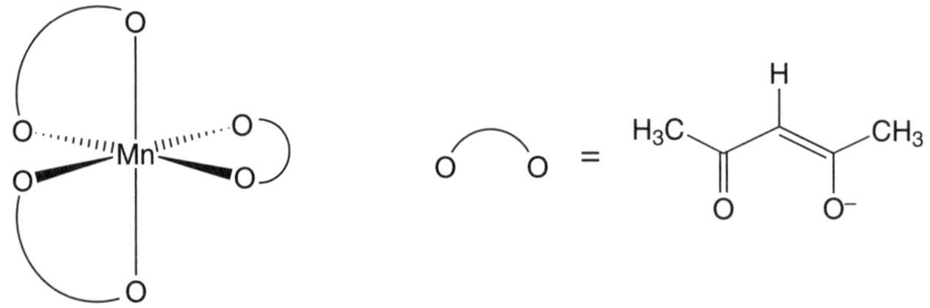

This Mn^{III} complex has four d electrons, and in this experiment you will determine whether the compound is high or low spin.

Magnetic Susceptibility

One of the purposes of this experiment is to acquaint you with the experimental methods used to measure magnetic susceptibilities. When a substance is placed in an external magnetic field, the substance will produce its own field, which will add to (paramagnetism) or subtract from (diamagnetism) the applied field. If the external magnetic field is relatively weak, this induced magnetic field inside the substance, called the magnetization I, is proportional to the external magnetic field H:

$$I = \kappa H \tag{1}$$

The proportionality constant κ, which is a measure of the tendency or suscepti-bility of the substance to interact with an applied field, is called the volume magnetic susceptibility. This volume susceptibility is usually converted to susceptibility per gram of substance, χ, by dividing by the density of the substance, d:

$$\chi = \frac{\kappa}{d} \tag{2}$$

The magnetic susceptibility per mole of substance, χ_M, is of most value for chemical applications. It is obtained by multiplying χ by the molecular weight, M, of the compound.

$$\chi_M = (\chi)(M) \tag{3}$$

The molar susceptibility, χ_M, is positive if the substance is paramagnetic and negative if it is diamagnetic.

Because the overall χ_M of a compound is the sum of the susceptibilities of the paramagnetic unpaired electrons and of the diamagnetic paired electrons, the susceptibility of only the unpaired electrons, χ_M', may be obtained from the additive relationship

$$\chi_M = \chi_M' + \chi_M \text{ (metal core electrons)} + \chi_M \text{ (ligands)} + \chi_M \text{ (other)} \tag{4}$$

In this equation, the last three terms represent, respectively, the susceptibilities of the diamagnetic paired electrons of the inner core of the metal, the ligands, and any other ions or molecules that may be present in the solid. Thus, in order to determine the magnetic susceptibility of the unpaired electrons in the complex, it is necessary to correct the measured χ_M for the diamagnetism of the other groups. It is known that susceptibilities of diamagnetic groups change very little with their environment, hence, it is possible to calculate the diamagnetism of a molecule by adding together the diamagnetic contributions from each of the structural components of the molecule. Some diamagnetic corrections for ligands and ions are given in Table 12-1. The corrections for transition metal ions have only been estimated and frequently only the sum $\chi_M' + \chi_M$ (metal core electrons) is evaluated.

The paramagnetic susceptibility of a substance is not constant but varies with temperature. For a substance consisting of noninteracting paramagnetic centers (as in coordination compounds, in which ligands insulate the metal ions from each other), the paramagnetic susceptibility often depends on the absolute temperature of the substance according to the Curie law:

$$\chi_M' = \frac{C}{T} \tag{5}$$

where C is a constant called the Curie constant and T is the absolute temperature (kelvin). In many other compounds, however, the spins are not completely independent but are weakly influenced by those of neighboring molecules. In this case, the paramagnetic susceptibility does not obey the Curie law but instead follows the Curie–Weiss law,

$$\chi_M' = \frac{C}{(T - \theta)} \tag{6}$$

Table 12-1
Diamagnetic Corrections for Ligands and Ions[a]

Formula	$10^6\,\chi_M(\text{cm}^3\,\text{mol}^{-1})$	Formula	$10^6\,\chi_M(\text{cm}^3\,\text{mol}^{-1})$
$acac^-$	-52	ClO_4^-	-32
Ag^+	-28	en^b	-46
BF_4^-	-37	F^-	-9
Ba^{2+}	-26	H_2O	-13
Br^-	-35	I^-	-51
Ca^{2+}	-10	K^+	-15
CN^-	-13	Mg^{2+}	-5
NCS^-	-31	NH_3	-18
CO	-10	NO_3^-	-19
CO_3^{2-}	-28	Na^+	-7
$CH_3CO_2^-$	-30	OH^-	-12
C_5H_5N	-49	Pb^{2+}	-32
$C_5H_5^-$	-65	SO_4^{2-}	-40
Cl^-	-23	Zn^{2+}	-15

[a] The calculated inner core diamagnetism of the first-row transition metals is approximately $-13 \times 10^6\,\text{cm}^3\,\text{mol}^{-1}$.
[b] Ethylenediamine is abbreviated en.

where θ is the Weiss constant. Experimentally, the values of C and θ are determined by measuring χ_M' at a variety of temperatures, and plotting $1/\chi_M'$ versus T. The resulting straight line has a slope of $1/C$ and intersects the x axis at $T = \theta$ (Fig. 12-2):

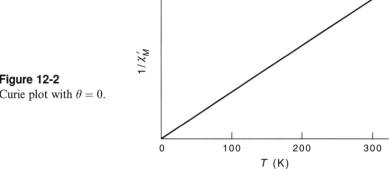

Figure 12-2
Curie plot with $\theta = 0$.

If only the spin angular momentum of each unpaired electron (and not its orbital angular momentum) contributes to χ_M', then C is related to the number of unpaired electrons on each paramagnetic center, n, by the equation:

$$C = \tfrac{1}{8}n(n+2) \qquad (7)$$

where C is expressed in units of cubic centimeters kelvin per mole ($\text{cm}^3\,\text{K}\,\text{mol}^{-1}$). This expression is very useful, because it permits you to determine the number of unpaired electrons in a paramagnetic compound from the slope of the line obtained by plotting $1/\chi_M'$ versus T.

In order to calculate the correct value for n, it is therefore necessary to determine the temperature dependence of χ'_M. Because a temperature dependence study requires more sophisticated equipment than will be used in this experiment, we will assume that χ'_M follows the Curie law, that is, θ is very small compared to the temperature at which the compound is studied. This assumption will not introduce a large error, because the magnitude of θ is normally less than 30 K. By making this assumption, Eqs. 5 and 7 combine to give

$$\chi'_M T = \tfrac{1}{8} n (n + 2) \tag{8}$$

where $\chi'_M T$ is expressed in units of cubic centimeters kelvin per mole (cm³ K mol⁻¹). In this experiment, the value of χ'_M will be measured, and Eq. 8 will be used to compute n for Mn(acac)₃.

For paramagnetic substances, chemists sometimes compute a related number, the effective magnetic moment or μ_{eff}, which is defined by the equation:

$$\mu_{\text{eff}} = \sqrt{8\chi'_M T} \tag{9}$$

where χ'_M is expressed in units of cubic centimeters per mole (cm³ mol⁻¹) and μ_{eff} has units of Bohr magnetons (μ_B). For compounds that obey the Curie law, μ_{eff} is a constant that is independent of temperature.

There are a number of methods of determining χ_M, some requiring expensive and delicate equipment. We will describe two of the simplest methods to measure χ_M: the Gouy method and the Evans method. Your instructor will tell you which method will be employed in this experiment.

The Gouy Method
A schematic drawing of the equipment used in the Gouy method is shown in Figure 12-3. Details of construction are given in the references at the end of the

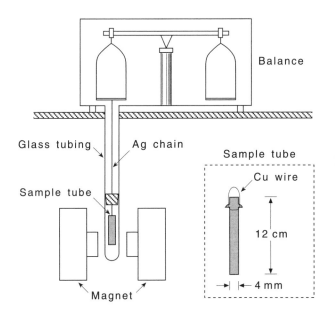

Figure 12-3
Equipment used in the Gouy method.

experiment. Although the components may vary considerably, a few guidelines might be mentioned. The sensitivity of the balance must be at least 0.1 mg, although this depends somewhat on the size of the sample and the strength of the magnet. A chain made of silver (or other non-magnetic metal) is suspended from the left-hand pan of the balance through a hole in the floor of the balance case and a hole in the table. The sample tube (Fig. 12-3) is attached to the silver chain with a fine copper wire and allowed to hang in the magnetic field, with the bottom of the sample tube near the center of the pole faces of the magnet. The silver chain and sample tube are enclosed in glass to prevent drafts from disturbing the weighings. The magnet may be either permanent or an electromagnet that generates a magnetic field of at least 5000 G, where G = gauss. Some provision must be made to remove the magnetic field from the sample, so that the sample may be weighed in the absence and presence of the magnetic field. With an electromagnet, the field may be simply turned off. With a permanent magnet, the magnet must be physically moved from the sample area.

To determine the magnetic susceptibility of a substance, one first calibrates the Gouy apparatus by using a standard compound whose χ_M is known. This calibration involves measuring the change in weight of a given mass of standard in the presence and absence of the magnetic field. Once this measurement has been made, then the change in weight, in and out of the field, of an unknown compound can be used to calculate its susceptibility.

There is a variation of the Gouy method in which the sample is kept motionless and the *magnet* is weighed (Fig. 12-4). This "inverse Gouy method," which was invented by the chemist Dennis Evans, uses much simpler and much smaller equipment. Two small C-shaped magnets are joined back-to-back to a slender wire, and the wire is strung horizontally between two posts to form a torsion assembly. Between the poles of one of the two magnets (called the restoring

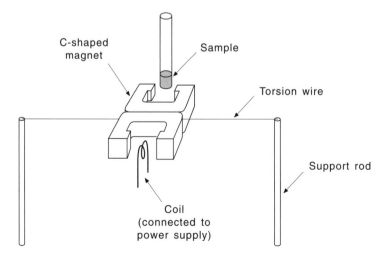

Figure 12-4
Schematic of the inverse Gouy method, in which the magnet is weighed instead of the sample.

magnet) is placed a small electrical coil; the gap between the poles of the other magnet (called the analyzing magnet) is left open to receive the sample. The sample is placed in a glass tube. When the tube is inserted into the gap of the analyzing magnet, the attraction (or repulsion) between the sample and this magnet causes the torsion assembly to twist about the slender wire. If a current is passed through the electrical coil in the gap of the restoring magnet, the torsion assembly can be returned to its original position. The current necessary to do this is a measure of the force between the sample and the analyzing magnet. The magnetic susceptibility of the sample can be calculated once the apparatus has been calibrated with a standard. The equipment necessary to carry out such a measurement can fit in a small box, and is commercially available at a reasonable price.

The Evans Method
Another way to measure χ_M is by means of nuclear magnetic resonance (NMR) spectroscopy. In this method, which was also invented by Dennis Evans, the compound is dissolved in a solvent that gives a sharp singlet in its ^{1}H NMR spectrum. This solution is placed into an NMR tube along with a sealed capillary that contains the pure solvent. The paramagnetic compound causes the solvent resonance to shift away from its normal position; as a result the NMR spectrum will contain two resonances, one due to the solution and the other to the pure solvent in the capillary. The frequency difference is related to the magnetic susceptibility of the solute (in units of cm^3 mol^{-1}) by the following formula:

$$\chi_M = (477)\frac{\Delta v}{Q v_{\mathrm{I}} c} \tag{10}$$

where Δv is the frequency difference in hertz between the shifted resonance and the pure solvent reference peak, v_{I} is the frequency in hertz of the radio waves generated by the NMR instrument, and c is the concentration of solute in moles per liter. The factor Q in Eq. 10 depends on the type of magnet used in the NMR spectrometer: If the system uses an electromagnet, then $Q = 1$; if the system uses a superconducting magnet, then $Q = 2$. The difference relates to the different magnet design and the orientation of the magnetic field relative to the spinning axis of the NMR sample.

EXPERIMENTAL PROCEDURE

Safety note: Potassium permanganate is a strong oxidant. If you spill some on your skin, wash it off immediately with large amounts of water.

Tris(acetylacetonato)manganese(III), Mn(acac)$_3$
In a 250-mL beaker, prepare a solution of 3.75 g (0.027 mol) of potassium permanganate, KMnO$_4$, and 75 mL of distilled H$_2$O. Warm the stirred solution to 80 °C on a hot plate to dissolve all the solid. Then cool the solution to room temperature by packing the beaker in ice. When the solution has cooled, stir it rapidly and slowly add 17 mL (16.6 g, 0.166 mol) of acetylacetone in several ali-

quots over a few minutes. (If you add the acetylacetone too quickly, the solution will generate large amounts of foam that may overflow the flask.) Note the color change. After the addition is complete, boil the solution for 5 min, and then chill the beaker in ice. Collect the shiny brown-black crystals on a coarse fritted filter (see Figure 13-1) and wash them three times with 10-mL portions of distilled water. Dry the crystals thoroughly by pulling air through the frit for at least 10 min. If possible, dry the crystals under vacuum for at least 30 min.

Measurement of Magnetic Susceptibility by the Evans Method

Weigh a clean and dry NMR tube (5-mm outer diameter) on an analytical balance to the nearest 0.1 mg (or to the nearest 0.01 mg if possible). Add between 2 and 5 mg of $Mn(acac)_3$ to the NMR tube; this amount of $Mn(acac)_3$ is less than what is needed to cover the bottom of the tube. Reweigh the NMR tube to determine exactly the amount of solid added.

If you will be using a NMR spectrometer equipped with an electromagnet, carry out the following procedure. Into the NMR tube, place a sealed capillary that contains pure chloroform, $CHCl_3$ (the capillaries can be made by syringing $CHCl_3$ into a melting point capillary, and then sealing the open end in a flame). Make sure to insert the capillary so that the end filled with solvent is resting on the bottom of the NMR tube. With a syringe, add exactly 0.70 mL of $CHCl_3$ to the NMR tube. Cap the tube and shake it gently to dissolve the $Mn(acac)_3$ completely. When the solid has completely dissolved, record the NMR spectrum of the solution (see your instructor for directions concerning the use of the NMR instrument). There should be two peaks near δ 7: One of the two peaks is due to the $CHCl_3$ in the capillary and the other due to the $CHCl_3$ that has been paramagnetically shifted by the dissolved $Mn(acac)_3$. Determine the chemical shifts of these two peaks, and measure the separation between them in hertz. For example, if the spectrometer frequency is 100 MHz and the two peaks are 0.3 ppm apart, then the separation is $0.3 \times 100 = 30$ Hz. The magnetic susceptibility of your sample can be determined from the weight of your sample, the volume of $CHCl_3$ used, the chemical shift difference Δv between the two peaks, and radio frequency used. Use Eq. 10 with $Q = 1$.

If you will be using a NMR spectrometer equipped with a superconducting magnet, carry out the same procedure except use deuterated chloroform ($CDCl_3$) both in the capillary and as the solvent to dissolve the $Mn(acac)_3$. The measurement is conducted in the same way; the only difference is that the two peaks near δ 7 are due to the small (about 0.1%) residual amounts of $CHCl_3$ in the $CDCl_3$ solvent. Determine the magnetic susceptibility of your sample from Eq. 10 with $Q = 2$.

Measurement of Magnetic Susceptibility by the Gouy Method

Scratch a horizontal line on the Gouy tube about 2 cm from the top, if this has not already been done. For the weight measurements, the tube should always be filled to this line. To remove paramagnetic impurities from the tube, clean it with Nochromix cleaning solution (do not use chromic acid, which will add paramagnetic impurities!). Thoroughly rinse the tube with water and acetone, and dry it in the oven. Do not wipe the tube with a dry towel; this gives the tube a static charge that significantly affects the weighings.

Weigh the empty tube on the chain with the field off. With the field on, weigh the tube again. Although "pure glass" is diamagnetic, paramagnetic impurities may cause the tube to be attracted by the field rather than repelled. The difference between these two weighings, on minus off, is δ and will be used to correct for the magnetism of the tube when the sample is weighed in the tube. It is necessary to maintain the same magnetic field for all measurements when the field is on. With electromagnets this will require a constant electric current (and therefore field). Because the current will decrease as the coils begin to heat, the current will have to be adjusted frequently. Current regulators are available that will conveniently provide this control.

To correct for the magnetism of air when it is displaced by the sample, the volume occupied by the air must be determined. Fill the tube to the line with water, and then weigh it with the field off. From the weight and known density, d, of water at the existing temperature, the volume (V) may be calculated. The volume susceptibility of air is 0.029×10^{-6}.

To determine the calibration constant for the apparatus, fill the dry tube to the line with the solid standard. The largest source of error in the Gouy method is inhomogeneously packed sample tubes. To minimize this problem, the sample should be finely powdered (use a mortar and pestle) and introduced into the tube in small portions. After each addition, firmly tap the tube on a hard surface. Careful packing of the tube will require 20–30 min. Weigh the tube with the magnet off and again with the magnet on, using the same current as used previously. After each weighing of a solid sample with the field on, measure the temperature between the poles of the magnet. The difference in these two weighings (on minus off) is designated Δ and is a measure of the magnetic susceptibility of both the sample and the tube. From the magnetic susceptibility per gram of the standard (χ), the mass in grams of the standard (m), and the values of δ, Δ, and V, the calibration constant (β) may be calculated

$$(\chi)(m) - (0.029 \times 10^{-6})V = \beta(\Delta - \delta) \tag{11}$$

For calibration, either HgCo(NCS)₄ or [Ni(en)₃]S₂O₃, where en = NH₂CH₂CH₂NH₂, has proved to be very satisfactory. These compounds may be prepared easily in high purity, are stable, are not hygroscopic, and pack very well. The susceptibility per gram of HgCo(NCS)₄ at 20 °C is $\chi = 16.44 \times 10^{-6}$ cm³ g⁻¹. The susceptibility obeys the Curie–Weiss law with $\theta = -10$ K. The relatively high susceptibility of this compound sometimes causes the sample tube to cling to one of the poles of the magnet. This problem can be avoided by carefully positioning the sample tube midway between the poles. On the other hand, [Ni(en)₃]S₂O₃ is rarely drawn toward a pole because of its lower susceptibility. Its value of χ at 20 °C is 11.03×10^{-6} cm³ g⁻¹. It, too, obeys the Curie–Weiss law with $\theta = 43$ K.

After the value of β is obtained, the sample tube is cleaned and dried. The same procedure is repeated for the determination of the unknown, Mn(acac)₃.

The empty tube is weighed with the field on and off to obtain δ. Then the tube is carefully packed with finely powdered $Mn(acac)_3$. The filled tube is then weighed with the field on and off to obtain Δ. These measurements will then permit the calculation of χ and the molar susceptibility, χ_M.

Summarized below are the measurements that must be made first on the standard and then on the unknown:

A. Weight of empty tube, field off   ————————— g
B. Weight of empty tube, field on   ————————— g
C. Weight of tube filled to line with water, field off ————————— g
D. Weight of tube filled to line with solid, field off   ————————— g
E. Weight of tube filled to line with solid, field on   ————————— g
F. Temperature during measurements above   ————————— °C

As given, these weights are related to the terms in Eq. 11 in the following manner:

$V = (C - A)/d$, where d is density of water $(g\ mL^{-1})$ at ambient temperature

$\delta = B - A$

$\Delta = E - D$

$m = D - A$

To determine the reproducibility of your value, repeat the evaluation of χ for $Mn(acac)_3$ at least one more time by emptying and repacking the tube and then making the necessary weighings.

REPORT

Include the following:
1. Values of χ for $HgCo(NCS)_4$ and/or $[Ni(en)_3]S_2O_3$ at the temperature of your weighings, and the calibration constant β, if these were measured.
2. Data necessary to calculate χ, χ_M, χ'_M, and μ_{eff} for $Mn(acac)_3$.
3. Sample calculation of each of the above quantities, as well as the number of unpaired electrons in the complex.
4. Account for your experimentally determined number of unpaired electrons in terms of the electronic structure of $Mn(acac)_3$.

PROBLEMS

1. Account for the increase in the values of the diamagnetic corrections (Table 12-1) in going from the top to the bottom of any group in the periodic table.
2. By using the known values of χ for $HgCo(NCS)_4$ and $[Ni(en)_3]S_2O_3$, calculate the number of unpaired electrons in each of these complexes, and interpret the results in terms of the electronic structures of the complexes.
3. An iron(II) complex of the formula FeL_5^{2+} might have either square pyramidal or trigonal bipyramidal geometry. Would it be possible to differentiate between these geometries on the basis of a magnetic susceptibility

measurement of the complex? How would the number of unpaired electrons differ in the two geometries? (Use crystal field theory to show how the d orbitals split for the two structures.)

4. Make the assumption that only the spin angular momentum of the electron (and not its orbital angular momentum) contributes to χ'_M. Calculate the "spin-only" values of μ_{eff} for n equal to 0, 1, 2, 3, 4, and 5.

5. Rationalize Eq. 11.

6. The volume susceptibility of air is 0.029×10^{-6}. What causes its susceptibility to be positive?

7. In a Gouy measurement, if the standard were packed more compactly than the unknown, but both were packed evenly in the tube, how would this affect the calculated value of χ?

8. Why should chromic acid not be used to clean Gouy tubes?

9. A student measured the effective magnetic moment of NiCl$_4{}^{2-}$ and obtained a value of 3.2 μ_B. A value of 0.0 μ_B was obtained for Ni(CN)$_4{}^{2-}$. Rationalize these values in terms of the structures and crystal field splitting diagrams of these ions.

10. When the bottom of the tube is at the center of the pole faces, a Gouy tube containing HgCo(NCS)$_4$ has a certain weight in the magnetic field. If the magnet is raised so that the bottom of the tube is slightly below the center of the pole faces, will the observed weight be more or less than in the original measurement? Explain.

INDEPENDENT STUDIES

A. Prepare and determine the effective magnetic moment and number of unpaired electrons in Cr(acac)$_3$ (Fernelius, W. C.; Blanch, J. E. *Inorg. Synth.* **1957**, *5*, 130), the lanthanide complex Gd(thd)$_3$, where thd is 2,2,6,6-tetramethylheptane–3,5–dionate (Eisentraut, K. J.; Sievers, R. E. *Inorg. Synth.* **1968**, *11*, 94), or other compounds.

B. Using the Evans solution NMR method, determine the effective magnetic moment and number of unpaired electrons in the spin cross-over complex Fe[S$_2$CN(CH$_3$)$_2$]$_3$ as a function of temperature. (Hughes, J. G.; Lawson, P. J. *J. Chem. Educ.* **1987**, *64*, 973.)

C. Prepare the solid Gd$_3$Fe$_3$O$_{12}$ and study its magnetic properties. (Geselbracht, M. J.; Cappellari, A. M.; Ellis, A. B.; Rzeznik, M. A.; Johnson, B. J. *J. Chem. Educ.* **1994**, *71*, 696.)

D. Study the magnetic properties of various manganese oxides. (Teweldemedhin, Z. S.; Fuller, R. L.; Greenblatt, M. *J. Chem. Educ.* **1996**, *73*, 906.)

REFERENCES

Metal β-Diketonate Complexes

Bhattacharjee, M. N.; Chaudhuri, M. K.; Khathing, D. T. *J. Chem. Soc., Dalton Trans.* **1982**, 669. Preparation of Mn(acac)$_3$.

Mehrotra, R. C. *Metal β-Diketonates and Allied Derivatives*, Academic: New York, 1978.

Siedle, A. R. in *Comprehensive Coordination Chemistry*, Wilkinson, G.; Gillard, R. D.; McCleverty, J. A., Eds, Pergamon: New York, 1988, Chapter 15.4. Summary of β-diketonate complexes.

Magnetochemistry—Techniques and Interpretation

Carlin, R. *Magnetochemistry*, Springer-Verlag: New York, 1986. A specialist text on magnetism.

Figgis, B. N.; Lewis, J. in *Technique of Inorganic Chemistry*, Vol. IV, Jonassen, H. B.; Weissberger, A., Eds., Wiley: New York, 1965, p 137. Review of instrumental methods for measuring magnetic properties.

Jiles, D. *Introduction to Magnetism and Magnetic Materials: Properties*, Chapman and Hall: London, 1991.

Kahn, O. *Molecular Magnetism*, VCH: New York, 1993. An advanced text but with very readable explanations.

Quickenden, T. I.; Marshall, R. C. *J. Chem. Educ.* **1972**, *49*, 114. A primer on the units used in magnetism.

White, R. M. *Science* **1985**, *229*, 11. A review of the applications of magnetic solids.

Wood, R. *Understanding Magnetism*, Tab Books Inc.: Blue Ridge Summit, PA, 1988. An excellent general reference.

The Evans Method for Measuring Susceptibilites of Solutions

Evans, D. F. *J. Chem. Soc.* **1959**, 2003. The original paper on measuring susceptibilities by NMR spectroscopy.

Ostfeld, D.; Cohen, I. A. *J. Chem. Educ.* **1972**, *49*, 829. A good summary of the Evans method.

Sur, S. K. *J. Mag. Reson.* **1989**, *82*, 169. Evans method using high-field Fourier-transform spectrometers.

Preparation and Aquation of Co(NH$_3$)$_5$Cl^{2+}

Note: This experiment requires about 5 hours for Part A and 3 hours for Part B.

Of particular importance to the development of coordination chemistry are metal complexes of the type to be synthesized and characterized in this experiment. "Classical" coordination complexes of the transition metals typically involve unidentate ligands such as Cl$^-$, Br$^-$, I$^-$, NH$_3$, CN$^-$, and NO$_2$$^-$ and bidentate ligands such as ethylenediamine (H$_2$NCH$_2$CH$_2$NH$_2$), oxalate ($^-$O$_2$CCO$_2$$^-$), and carbonate (CO$_3$$^{2-}$).

Coordination compounds of Co^{III} and Cr^{III} have been of particular interest because their complexes undergo ligand exchange very slowly compared with complexes of many other transition metal ions. For example, Co(NH$_3$)$_6$$^{3+}$ and Cr(NH$_3$)$_6$$^{3+}$ are stable for months in acidic aqueous solutions; in contrast, the analogous compound Ni(NH$_3$)$_6$$^{2+}$ reacts virtually instantaneously under the same conditions to form Ni(OH$_2$)$_6$$^{2+}$. The study of complexes with slow ligand exchange rates laid the foundation of modern transition metal chemistry. In contrast, complexes with rapid ligand exchange rates are the key components of many important industrial and biological catalysts. Today, the rates of ligand exchange can be qualitatively accounted for by ligand field theory and molecular orbital theory.

The structures of the octahedral Co^{III} complexes that you will prepare are given below.

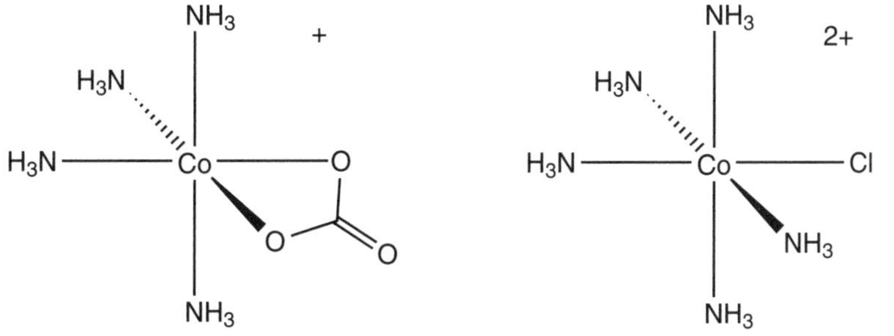

Both of these complexes are ionic species: They contain cations and anions. One important method of characterizing ionic substances is the determination of the ability of their solutions to conduct an electric current. Those substances whose solutions have the highest conductivity consist of the greatest number of ions. Thus, a 1 M solution of $[Co(NH_3)_4CO_3]NO_3$ will have a lower conductance than a solution of $[Co(NH_3)_5Cl]Cl_2$ of the same concentration. By measuring the conductivity of a solution of a compound, it is possible to determine whether a formula unit of that compound consists of 2, 3, 4, or more ions. Although measurements will be done on aqueous solutions of the complexes, polar organic solvents such as nitrobenzene or acetonitrile can frequently be used to obtain the same information for ionic compounds that either are not very soluble in water or react with water.

The synthesis of $[Co(NH_3)_4CO_3]NO_3$ will be carried out according to the equation,

$$Co^{2+} + 4\,NH_3(aq) + CO_3^{2-} + \tfrac{1}{2}\,H_2O_2 \longrightarrow [Co(NH_3)_4CO_3]^+ + OH^- \quad (1)$$

A convenient source of Co^{2+} is the salt $Co(NO_3)_2\cdot 6\,H_2O$, which is a coordination compound having the ionic formulation $[Co(OH_2)_6](NO_3)_2$. Because Co^{II} complexes, like those of Ni^{II}, react very rapidly by ligand exchange, the first step in the reaction might be expected to be $Co(OH_2)_6^{2+} + 4\,NH_3 + CO_3^{2-} \longrightarrow Co(NH_3)_4CO_3 + 6\,H_2O$. The latter Co^{II} complex could then be oxidized by the transfer of an electron to H_2O_2 to give the relatively unreactive Co^{III} ion, $Co(NH_3)_4CO_3^+$.

The preparation of $Co(NH_3)_5Cl^{2+}$ from the carbonato complex proceeds according to the following series of equations:

$$Co(NH_3)_4CO_3^+ + 2\,HCl \longrightarrow Co(NH_3)_4(OH_2)Cl^{2+} + CO_2 + Cl^-$$

$$Co(NH_3)_4(OH_2)Cl^{2+} + NH_3 \longrightarrow Co(NH_3)_5(OH_2)^{3+} + Cl^-$$

$$Co(NH_3)_5(OH_2)^{3+} + HCl \longrightarrow Co(NH_3)_5Cl^{2+} + H_2O$$

The first reaction in the preceding sequence probably occurs by means of the following mechanism:

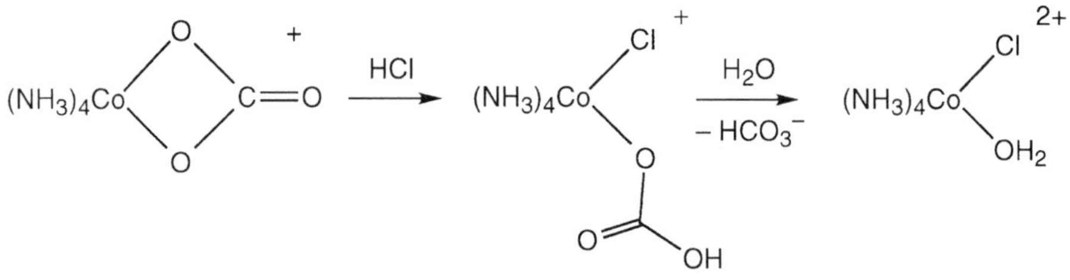

That O–C bond fission occurs in the intermediate has been established from ^{18}O isotopic exchange studies in several similar reactions of carbonato complexes. The subsequent steps in this preparation involve the substitution of one ligand in the coordination sphere by another. At first glance, one might expect these reactions to proceed according to S_N1 or S_N2 mechanisms, but even now there is considerable debate as to how these substitutions actually proceed. In Part B of this experiment, you will have an opportunity to postulate a mechanism for the reverse of the last reaction in the preceding series, the conversion of $Co(NH_3)_5Cl^{2+}$ to $Co(NH_3)_5(OH_2)^{3+}$.

Electrical Conductance of Solutions of Ionic Compounds

The determination of the number of ions constituting a given substance is largely a matter of defining conductance and then comparing conductances of the unknown ionic substances with those of known compounds. The definitions usually begin with resistance, since this is the quantity that is experimentally measured. The specific resistance, ρ, is defined as the resistance in ohms of a solution in a cell that has 1-cm^2 electrodes that are separated from each other by a distance of 1 cm. The reciprocal of ρ is the specific conductance, L. The resistance, R, of the same solution in a cell of nonstandard dimensions is obtained by multiplying ρ by a correction factor k, which depends on the geometry of the cell:

$$R = k\rho \qquad (2)$$

Experimentally, k is evaluated by measuring R for a given solution whose ρ has been measured in a standard cell and then calculating k from the preceding expression.

Because $\rho = 1/L$, Eq. 2 is usually expressed in terms of measured resistance, R, and the specific conductance:

$$R = \frac{k}{L} \qquad (3)$$

The cell constant, k, is frequently obtained from Eq. 3 by measuring the resistance, R, of a 0.02 M KCl solution whose specific conductance at 25 °C is 0.002768 ohm^{-1}. After k for the cell used in the study is determined, the measurement of R will allow the calculation of the specific conductance of any solution. To determine the conductance of a solution of an electrolyte, it is desirable to compare conductances under standard conditions. Thus, the molar conductance, Λ_M, is defined as the conductance of a 1-cm^3 cube of solution that contains 1 mol (or formula weight) of solute. Because the specific conductance, L, is the conductance of a 1-cm^3 cube of solution, the conductance per mole of solute may be calculated by dividing L by the number of moles present in 1 cm^3 of solution:

$$\Lambda_M = \frac{1000 \, L}{M} \qquad (4)$$

where M = molarity of the solution.

Comparisons of molar conductances with those of known ionic substances allow one to determine the number of ions present in a given salt. General ranges of Λ_M for 2, 3, 4, and 5 ion conductors at 25 °C in water solvent are tabulated as follows:

Number of Ions	Λ_M (ohm^{-1} cm^2 mol^{-1})
2	118–131
3	235–273
4	408–435
5	~560

Molar conductances of 2, 3, 4, and 5 ion conductors in other solvents at 25 °C are given in Appendix 2.

Ligand Exchange Kinetics

In the second part of this experiment, you will investigate the kinetics of a reaction involving the substitution of Cl$^-$ by H$_2$O in acidic solution

$$Co(NH_3)_5Cl^{2+} + H_2O \xrightarrow{H^+} Co(NH_3)_5(OH_2)^{3+} + Cl^-$$

Substitution reactions may proceed by a variety of mechanisms. One possibility is an S$_N$1 mechanism in which the rate-determining step is the breaking of the Co–Cl bond. The resulting open coordination site is then rapidly filled by a molecule of H$_2$O. Another possibility is an S$_N$2 mechanism in which the rate-determining step involves the attack of an H$_2$O molecule on the Co^{III} complex to form a short-lived seven-coordinate intermediate, which rapidly loses Cl$^-$ to generate the product.

Although these two mechanisms are very different, experimentally it is impossible to differentiate between them if the reaction is conducted in H$_2$O solvent, as the following equations show. The S$_N$1 mechanism predicts the first-order rate law,

$$\text{Rate of reaction of } Co(NH_3)_5Cl^{2+} = k_1[Co(NH_3)_5Cl^{2+}]$$

where k_1 is the first-order rate constant expressed in units of reciprocal seconds (s^{-1}). The S$_N$2 mechanism requires the overall second-order rate law,

$$\text{Rate of reaction of } Co(NH_3)_5Cl^{2+} = k_2[Co(NH_3)_5Cl^{2+}][H_2O]$$

where k_2 is the second-order rate constant (in M^{-1} s^{-1}). If H$_2$O is the solvent, however, the H$_2$O is in such large excess compared with the Co(NH$_3$)$_5$Cl^{2+} concentration, and so little of the H$_2$O is consumed in the actual reaction, that the H$_2$O concentration, [H$_2$O], remains essentially unchanged during the progress of the reaction. Consequently, one would predict the experimentally observed rate law for the S$_N$2 mechanism to be

$$\text{Rate} = k_{obs}[Co(NH_3)_5Cl^{2+}]$$

where $k_{obs} = k_2[H_2O]$. Thus, with H_2O as the solvent, the experimentally observed rate laws for the S_N1 and S_N2 reactions have the same mathematical form, even though the mechanisms are quite different. Experimentally, the reaction appears to obey a first-order rate law in both cases. Distinguishing S_N1 from S_N2 mechanisms must be done by other means.

A third possible ligand exchange mechanism is the acid-catalyzed aquation reaction. An example of this mechanism is found in the following reaction:

$$Co(NH_3)_5F^{2+} + H_2O \xrightarrow{H^+} Co(NH_3)_5(OH_2)^{3+} + F^-$$

The experimentally determined rate law is

$$\text{Rate of reaction of } Co(NH_3)_5F^{2+} = k[Co(NH_3)_5F^{2+}][H^+]$$

The acid catalysis apparently results from the addition of H^+ to the coordinated F^- ligand

$$(NH_3)_5Co-F^{2+} + H^+ \underset{\longleftarrow}{\overset{fast}{\longrightarrow}} (NH_3)_5Co-FH^{3+} \qquad K = \frac{[CoFH^{3+}]}{[CoF^{2+}][H^+]}$$

where this fast process is characterized by the equilibrium constant, K, shown. The HF group is a weaker ligand than F^-. One might visualize the proton as pulling off the F^- as HF, leaving an open coordination site, which is quickly occupied by an H_2O molecule.

$$(NH_3)_5Co-FH^{3+} + H_2O \xrightarrow[k_3]{} (NH_3)_5Co(OH_2)^{3+} + HF$$

If this last step is rate determining, this mechanism predicts the rate law

$$\text{Rate of reaction of } Co(NH_3)_5F^{2+} = k_3K[Co(NH_3)_5F^{2+}][H^+]$$

This expression is the same as the one observed experimentally. The only difference is that the experimental rate constant, k, is, in terms of this mechanism, equal to the product of equilibrium constant, K, and the rate constant for the rate-determining step, k_3:

$$k = k_3K$$

If one wishes to evaluate the rate constant k_3, one must determine the value of K by some other experimental means.

In this experiment, you will measure the rate of aquation of $Co(NH_3)_5Cl^{2+}$ at different H^+ concentrations and express your rate data in the form of a rate law. You will then postulate a mechanism that is consistent with your rate law. The ions $Co(NH_3)_5Cl^{2+}$ and $Co(NH_3)_5(OH_2)^{3+}$ have different extinction coefficients at 550 nm. The rate of conversion of one to the other can be determined by noting the change in the absorbance at that wavelength as a function of time.

EXPERIMENTAL PROCEDURE

Part A

Carbonatotetraamminecobalt(III) Nitrate, [Co(NH₃)₄CO₃]NO₃

Safety Note: Handle 30% H_2O_2 with rubber gloves; it can cause severe skin burns. In case of a spill, wash the affected area immediately with water. No precautions are necessary to protect the reaction mixtures from the atmosphere. Operations that necessitate heating of the solutions should be carried out in an efficient hood.

Dissolve 20 g (0.21 mol) of ammonium carbonate $(NH_4)_2CO_3$ in 60 mL of distilled water and add 60 mL of concentrated ammonium hydroxide, NH_4OH. While stirring, pour this solution into a solution containing 15 g (0.052 mol) of $[Co(OH_2)_6](NO_3)_2$ in 30 mL of H_2O. Then slowly add 8 mL of a 30% H_2O_2 solution. Pour the solution into a 250-mL beaker and concentrate the solution to 90–100 mL over a hot plate. Do not allow the solution to boil. While the solution is concentrating, add, in small portions, 5 g (0.05 mol) of $(NH_4)_2CO_3$. Filter the hot solution with a suction filtration apparatus (Fig. 13-1) and cool the filtrate in an ice bath. Under suction, filter off the red product crystals. Wash the

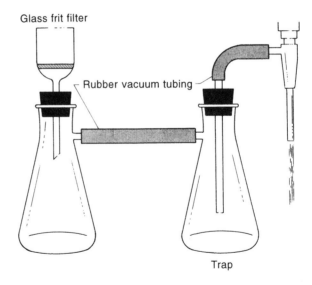

Figure 13-1
Suction filtration apparatus.

$[Co(NH_3)_4CO_3]NO_3$ in the filtration apparatus first with a few milliliters of water (the compound is somewhat soluble) and then with a similar amount of ethanol. Calculate the yield. (Save a portion of your product for the conductance determination.)

Chloropentaamminecobalt(III) Chloride, $[Co(NH_3)_5Cl]Cl_2$

Dissolve 5.0 g (0.02 mmol) of $[Co(NH_3)_4CO_3]NO_3$ in 50 mL of H_2O and add 5–10 mL of concentrated HCl until all of the CO_2 is expelled. Neutralize the reaction solution with concentrated aqueous NH_3 and then add about 5 mL excess (spot some of the solution on pH paper to check the pH). Heat for 20 min, again avoiding boiling; $[Co(NH_3)_5(OH_2)]^{3+}$ is formed. Cool the solution slightly and add 75 mL of concentrated HCl. Reheat for 20–30 min and observe the change in color. Cool the solution to room temperature, and decant the solution from the purple-red crystals of the product. Wash the compound several times, by decantation, with small amounts of ice-cold distilled water, then filter under a water aspirator vacuum with a glass fritted funnel (medium porosity). Wash with several milliliters of ethanol. Drying in an oven at 120 °C to remove solvent yields $[Co(NH_3)_5Cl]Cl_2$. Calculate the yield. (Do not discard the compound because part of it will be needed in the kinetics experiment.)

Conductivities of $[Co(NH_3)_4(CO_3)]NO_3$ and $[Co(NH_3)_5Cl]Cl_2$

You will measure the conductances of aqueous solutions of your coordination compounds in a conductance cell similar to that shown in Figure 13-2.

Carefully read the instructions provided by the manufacturer for the operation of the conductivity bridge. If you need help, ask the instructor. In making all resistance measurements, thermostat the cell containing the solution at 25 °C for

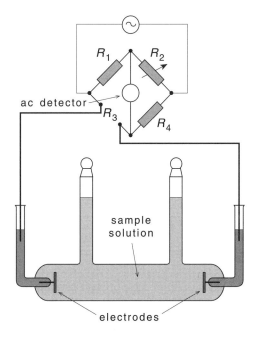

Figure 13-2

Conductance cell. Here R_1, R_2, and R_4 are resistance values in the conductance bridge, which is used to match the resistance, R_3, of the solution. (From Atkins, P. *Physical Chemistry*, 5th ed., Freeman: New York, 1994, p. 835.)

approximately 10 min before making a reading. Use distilled water in all solution preparations.

1. Prepare a 0.02 M KCl aqueous solution and obtain the cell constant, k, from the resistance measured for the solution.
2. Prepare 500 mL of 0.001 M aqueous solutions of $[Co(NH_3)_4CO_3]NO_3$ and $[Co(NH_3)_5Cl]Cl_2$ and measure their resistances. Make the measurements immediately after the solutions are prepared, because significant decomposition occurs on standing overnight.

Be certain to rinse the cell well with distilled water between measurements. When you have completed the experiment, rinse the cell thoroughly and leave it filled with water.

Calculate the molar conductances of the two Co^{III} complexes. These measurements are fairly sensitive tests of the ionic purity of your compounds.

Part B

Determination of the Aquation Rate of $[Co(NH_3)_5Cl]Cl_2$

Although some reactions are catalyzed by light and oxygen, this reaction is not, and precautions to eliminate them need not be taken. Before beginning the experiment, standardize the ultraviolet–visible (UV–vis) spectrophotometer (if necessary) by following the instructions provided by your instructor.

In a bath set at 60 °C, thermostat for at least 15 min four 100-mL volumetric flasks, one containing 0.05 M HNO$_3$, a second containing 0.10 M HNO$_3$, a third containing 0.30 M HNO$_3$, and a fourth containing 0.60 M HNO$_3$ (concd HNO$_3$ is 15.9 M; see Appendix 1). To each solution, add sufficient $[Co(NH_3)_5Cl]Cl_2$ to give a $1.2 \times 10^{-2} M$ complex concentration. Shake the flasks until all of the complex has dissolved, and then return the flasks to the thermostatted bath. When the solution has been kept in the bath for 15 min and has again reached a temperature of 60 °C, begin withdrawing samples with a pipet (use a rubber bulb, not your mouth!). At approximately 15-min intervals, withdraw 10-mL samples from each reaction mixture. Record the time of withdrawal, and measure the visible spectrum of each sample immediately after removing it from the thermostatted solution. Take eight aliquots from each reaction solution during the course of the rate study.

A variety of UV–vis spectrophotometers can be used to follow the progress of the reaction. If a scanning instrument is available, the absorption spectrum of each sample should be taken over a wavelength range of approximately 350–650 nm. All the spectra for each kinetic run should be plotted on the same piece of paper, and each curve should be labeled to ensure that it can later be assigned to a given solution and time of withdrawal. Only the changes in absorption at 550 nm will be used to determine the rate law, but it is of interest to note the other spectral changes that occur during the reaction. If a fixed wavelength spectrophotometer is available, the wavelength may be set at 550 nm and the absorbance at that wavelength may be recorded for each sample at a given time.

Use sample cells (see Appendix 3) of 1-cm path length for the measurements. When handling the cells, never touch the polished faces; fingerprints significantly reduce the transmission of light. Before installing a cell in the sample compartment, wipe the outside of the cell clean and dry with a soft tissue. Slowly scan the spectrum of the solution, or measure the absorbance (or transmittance) at 550 nm with a fixed wavelength instrument. After making the measurement, rinse the cell thoroughly with distilled water and acetone and then dry in the air or under a stream of N$_2$.

In kinetic studies of substitution reactions, it is usually valid to assume that the rate law will exhibit a first-order dependence on the complex concentration. Both S$_N$1 and S$_N$2 mechanisms predict the rate law, rate $= k_{obs}$[complex]. If the reaction is acid catalyzed, however, the rate law rate $= k$[complex][H$^+$]^n, where $n = 1, 2, 3$, and so on, will hold. Because H$^+$ is not consumed during the reaction, the [H$^+$]^n term is a constant in the rate law, which reduces to the expression: rate $= k_{obs}$[complex], where $k_{obs} = k$[H$^+$]^n. Thus, for all three possible mechanisms, k_{obs} can be obtained from a first-order kinetic plot of your data, and your task in this Experiment is to determine whether k_{obs} is dependent on the concentration of H$^+$.

From your absorbance and time data you will determine a first-order rate constant that will either be or not be dependent on the H$^+$ concentration. To calculate the rate constant from your measurements, you will need to know the absorbance at infinite time, A_∞. This value can be calculated from the known extinction coefficient of Co(NH$_3$)$_5$(OH$_2$)$^{3+}$, which is 21.0 cm$^{-1}M^{-1}$ at 550 nm. The first-order rate constant is equal to the slope of a plot of $\ln(A - A_\infty)$ versus time, t, where A is the absorbance of the solution at any given time, t.

REPORT

Include the following:
Part A
1. Percentage yields of [Co(NH$_3$)$_4$CO$_3$]NO$_3$ and [Co(NH$_3$)$_5$Cl]Cl$_2$.
2. Values of Λ_M for the preceding complexes and your conclusions as to the number of ions in each compound.

Part B
1. Absorbance measurements or spectra taken during the kinetic run.
2. Kinetic plots of data used to determine the rate constants.
3. Justification for using a plot of $\ln(A - A_\infty)$ versus t to obtain the first-order rate constants.
4. Rate constants and rate law.
5. Proposed mechanism.

PROBLEMS

Part A
1. Balance Eq. 1 for the preparation of [Co(NH$_3$)$_4$CO$_3$]NO$_3$.

2. How would you establish experimentally that the O–C bond, rather than the Co–O bond, is the one that is broken in the following reaction?

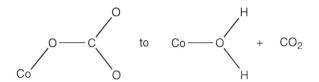

3. The conductance of an aqueous solution of $[Co(NH_3)_5Cl]Cl_2$ changes on standing overnight. Would you expect it to increase or decrease?

4. In the isolation of $[Co(NH_3)_4CO_3]NO_3$ and $[Co(NH_3)_5Cl]Cl_2$, why are the solids washed with ethanol after having first been washed with water?

5. Why are 500 mL of the 0.001 M $[Co(NH_3)_4CO_3]NO_3$ and $[Co(NH_3)_5Cl]Cl_2$ solutions prepared when only 50–100 mL are required for the conductance measurements?

6. How do $[Co(NH_3)_4CO_3]NO_3$ and $[Co(NH_3)_5CO_3]NO_3$ differ structurally? How would you experimentally distinguish between these two compounds?

7. Will the cell constant, k, change if the electrodes in a conductivity cell are bent or moved? Why?

Part B

1. Account for the fact that *trans*-$Co(NH_3)_4Cl_2^+$ undergoes substitution of one Cl^- by water at a rate that is 1000 times faster than that of $Co(NH_3)_5Cl^{2+}$. Both react by the same mechanism.

2. The aquation of $Co(NH_3)_5Cl^{2+}$ is accelerated by Ag^+. Propose a mechanism for this acceleration.

3. Suggest another method, other than spectrophotometry, of determining the rate of aquation of $Co(NH_3)_5Cl^{2+}$.

4. Why was 550 nm chosen as the wavelength at which the reaction was followed?

5. Consider the reaction:

$$Co(NH_3)_5Cl^{2+} + NH_3 \rightleftharpoons Co(NH_3)_6^{3+} + Cl^-$$

The rate law for this reaction may be written in the general form: rate $= k[Co(NH_3)_5Cl^{2+}]^x[NH_3]^y$, where the orders, x and y, are to be determined. You carry out rate studies with $[Co(NH_3)_5Cl^{2+}] = 0.001 M$ and several different NH_3 concentrations (e.g., $[NH_3] = 0.2, 0.3, 0.4,$ and 0.6 M). From the results of these kinetic runs, how would you determine x and y?

6. Why was HNO_3, not HCl, used in the aquation kinetics study?

INDEPENDENT STUDIES

A. Measure the infrared (IR) spectra of [Co(NH$_3$)$_4$(CO$_3$)]NO$_3$ and [Co(NH$_3$)$_5$Cl]Cl$_2$ as Nujol mulls (see Experiment 19 for details of the IR technique). Assign the absorptions characteristic of NH$_3$, CO$_3^{2-}$, and NO$_3^-$. How are the two IR spectra similar and how are they different?

B. Prepare and characterize other Co(NH$_3$)$_5$X^{2+} derivatives, where X$^-$ = F$^-$, Br$^-$, I$^-$, NO$_2^-$, NO$_3^-$, CO$_3^{2-}$, and so on. (Olson, G. L. *J. Chem. Educ.* **1969**, *46*, 508.)

C. Isolate and characterize *cis*-Co(NH$_3$)$_4$(OH$_2$)Cl^{2+}, which was formed as an intermediate in this experiment. (Kauffman, G. B.; Pinnell, R. P. *Inorg. Synth.* **1960**, *6*, 176.)

D. Exchange the protons of [Co(NH$_3$)$_5$Cl]Cl$_2$ with deuterium, using D$_2$O. Compare the IR spectrum of the resulting [Co(ND$_3$)$_5$Cl]Cl$_2$ with that of the protio form. (Sacconi, L.; Sabatini, A.; Gans, P. *Inorg. Chem.* **1964**, *3*, 1772.)

E. Determine the kinetics of nitrito to nitro linkage isomerism in Co(NH$_3$)$_5$(ONO)$^{2+}$. (Jackson, W. G. *J. Chem. Educ.* **1991**, *68*, 903.)

F. Determine rates of aquation of Co(NH$_3$)$_5$Cl^{2+} at two other temperatures spanning a total range of 30 °C. From the results, calculate the activation parameters (E_a, $\Delta H^{\ddagger}$, $\Delta S^{\ddagger}$) for the reaction. (See a kinetics text for those calculations.)

G. Prepare Co(NH$_3$)$_5$(SO$_3$)$^+$ and study its reaction with NO$_2^-$. (Richards, L. *J. Chem. Educ.* **1988**, *65*, 82.)

H. Study the rate of hydrolysis of Co(NH$_3$)$_5$Cl^{2+} in basic solution (OH$^-$) to form Co(NH$_3$)$_5$(OH)$^{2+}$. (Chan, S. C.; Hui, K. Y.; Miller, J.; Tsang, W. S. *J. Chem. Soc.* **1965**, 3207.)

I. Measure the equilibrium constant for the reaction:

$$\text{Co(NH}_3\text{)}_5\text{Cl}^{2+} \;+\; \text{H}_2\text{O} \;\longrightarrow\; \text{Co(NH}_3\text{)}_5\text{(OH}_2\text{)}^{3+} \;+\; \text{Cl}^-$$

(Taube, H. *J. Am. Chem. Soc.* **1960**, *82*, 524)

J. Study the trans to cis isomerization of [Co(NH$_3$)$_4$Cl$_2$]Cl. (Borer, L. L.; Erdman, H. W. *J. Chem. Educ.* **1994**, *71*, 332.)

REFERENCES

[Co(NH$_3$)$_4$CO$_3$]NO$_3$ and [Co(NH$_3$)$_5$Cl]Cl$_2$

Buckingham, D. A.; Clark, C. R. *Inorg. Chem.* **1994**, *33*, 6171. Kinetics of reactions of Co^{III} carbonate complexes.

Buckingham, D. A.; Clark, C. R. *Inorg. Chem.* **1993**, *32*, 5405. Acid-catalyzed hydrolysis of Co^{III} carbonate complexes.

Oldham, C. in *Comprehensive Coordination Chemistry*, Wilkinson, G.; Gillard, R. D.; McCleverty, J. A., Eds., Pergamon: New York, 1987; Vol. 2, Chap. 15.6.4. A review of metal carbonate complexes.

Schlessinger, G. G. *Inorg. Synth.* **1960**, *6*, 173. The preparation of [Co(NH$_3$)$_4$CO$_3$]NO$_3$.

Techniques

Any modern quantitative analysis text will discuss the fundamentals of absorption spectroscopy.

Boggess, R. K.; Zatko, D. A. *J. Chem. Educ.* **1975**, *52*, 649. Use of conductance data for structure determination of metal complexes.

Geary, W. J. *Coord. Chem. Rev.* **1971**, *7*, 81. Conductivities of electrolytes in nonaqueous solvents.

Kinetics and Reaction Mechanisms

Espenson, J. H. *Chemical Kinetics and Reaction Mechanisms*, 2nd ed., McGraw-Hill: New York, 1995.

Inorganic Reaction Mechanisms, Specialist Periodical Report, The Chemical Society, Burlington House: London. A series of annual volumes reviewing the literature in this area.

Jordan, R. B. *Reaction Mechanisms in Inorganic and Organometallic Systems*, 2nd ed., Oxford: New York, 1998.

Wilkins, R. G. *Kinetics and Mechanism of Reactions of Transition Metal Complexes*, 2nd ed., VCH: New York, 1991. A textbook including problems.

For General References to Coordination Chemistry, See Experiment 10

Optical Resolution of Co(en)$_3$$^{3+}$

Note: This experiment requires 6 hours spread over two laboratory periods.

One fascinating and important property of molecules is that they often exist in right- or left-hand forms. Such forms are called *enantiomers* or optical isomers, the latter term referring to the ability of such isomers to rotate the plane of polarized light. If equal amounts of the two enantiomers are present in a sample, the optical effects of the right- and left-hand forms cancel and no net rotation of light is seen. Such samples are called *racemic* mixtures. In the present experiment, you will synthesize a racemic mixture of a coordination complex, and then separate it into its two enantiomers.

Optical activity is frequently associated with organic molecules containing an asymmetric carbon atom, as, for example, in lactic acid.

Optical activity, however, is a far more general phenomenon and may be found in any molecule that cannot be superimposed upon its mirror image. In inorganic chemistry, there are many examples of optically active compounds that come in right- and left-hand forms. Although a few tetrahedral inorganic compounds have been resolved into their enantiomers, optical activity in octahedral transition metal complexes has been studied far more extensively.

Many octahedral complexes of transition metals can be resolved into two enantiomers. Some of the earliest work in this area was done in 1912 by Alfred Werner on Co(en)$_3$$^{3+}$ (where en $= NH_2CH_2CH_2NH_2$). The enantiomers of Co(en)$_3$$^{3+}$ that he resolved have the structures shown in Figure 14-1.

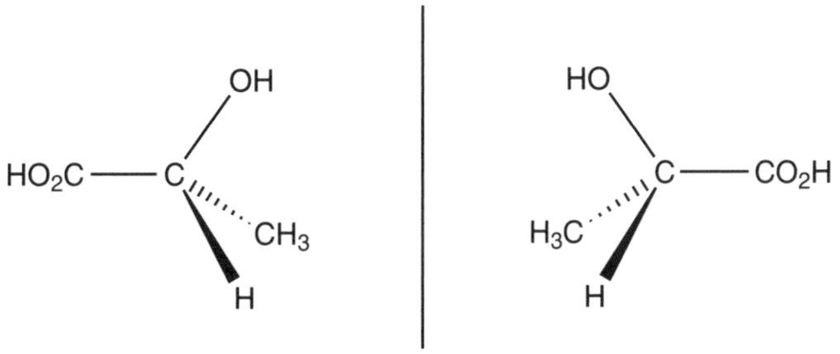

143

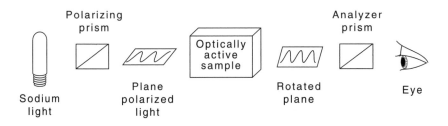

Λ (+) enantiomer Δ (−) enantiomer

Figure 14-1
Structures of the enantiomeric forms of $Co(en)_3{}^{3+}$.

One of the isomers rotates plane polarized light toward the right (dextrorotatory) while the other isomer rotates the light by the same amount in the opposite (levorotatory) direction. These directions are designated (+) and (−) (or sometimes d and l), respectively. Because of the availability of sodium arc lamps as a light source, light of 589.3-nm wavelength (the sodium D line) is frequently used in the determination of the rotations. Passing this light through a polarizing prism gives plane polarized light, whose electric field variation is shown in Figure 14-2.

Figure 14-2
Schematic of an optical polarimeter.

In order for the rotated polarized light to pass through the analyzer prism, this prism must be rotated, relative to the polarizing prism, to the right or left by an angle that is equal to the rotation caused by the sample. Thus, the direction and number of degrees of rotation (α) may be measured experimentally. As in any form of spectroscopy, the size of the rotation depends not only on the nature of the optically active material but also on the length, b (in decimeters), of the light path through the sample and the concentration, c (in g mL^{-1}), of the sample in a solvent. A parameter called the specific rotation $[\alpha]_\lambda$ has been defined as follows:

$$[\alpha]_\lambda = \frac{\alpha}{bc} \qquad (1)$$

The wavelength, λ, of light is also specified. If the rotation is measured at the wavelength of the sodium emission line at 589.3 nm (as is commonly done), the specific rotation is designated $[\alpha]_{589.3}$. A unit that is frequently of more value for comparison between compounds is the molecular rotation, $[M]_\lambda$.

$$[M]_\lambda = \frac{M[\alpha]_\lambda}{100}$$

Because M is the molecular weight of the substance, $[M]_\lambda$ is a relative measure of its rotatory power on a molecular basis.

The rotatory power of a substance varies with the wavelength of the light employed. Thus, it is possible for a molecule to be dextrorotatory toward light of 589.3-nm wavelength, but levorotatory at another wavelength. The values of $[M]_\lambda$ as a function of wavelength are shown for $(+)\text{Co(en)}_3^{3+}$ in Figure 14-3. This plot is called an optical rotatory dispersion (ORD) curve.

Because the values of $[M]_\lambda$ for enantiomers at any given wavelength are of the same magnitude but of opposite sign, the ORD curve of $(-)\text{Co(en)}_3^{3+}$ may be obtained by rotating the curve for $(+)\text{Co(en)}_3^{3+}$ by 180° around the 0° line in the figure.

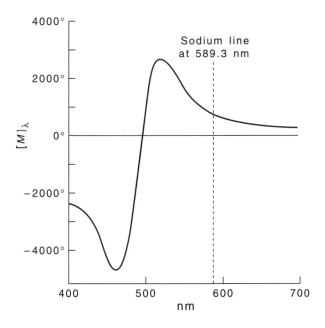

Figure 14-3
Optical rotatory dispersion curve for $\Lambda(+)\text{Co(en)}_3^{3+}$.

Although its optical activity indicated that $(+)Co(en)_3^{3+}$ must have one of the structures shown in Figure 14-1, the correct structure was not determined until 1954. By means of a special X-ray technique, a Japanese research group established the absolute configuration of $(+)Co(en)_3^{3+}$ as being that shown on the left in Figure 14-1. To show that the absolute configuration is known, the convention of using Λ to designate this configuration has generally been adopted. Its mirror image, $(-)Co(en)_3^{3+}$, must then have the absolute configuration shown on the right of Figure 14-1. It is labeled $\Delta(-)Co(en)_3^{3+}$. Although the lower case letters d and l indicate the nature of the optical absorption at a particular wavelength, and the upper case Greek letters Δ and Λ designate the handedness of the molecular structure, it is important to remember that these phenomena are not simply related. Molecules with Δ configurations can be levorotatory, for example.

Most biological compounds such as nutrients and drugs are optically active (and often only one enantiomer occurs in nature). In recent years, the study of optical activity in metal complexes has assumed great importance because optically active metal catalysts allow the selective synthesis of enantiomerically pure organic compounds. An example of an optically active metal catalyst is shown below:

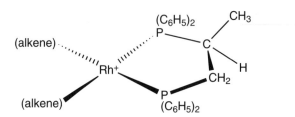

The preparation, resolution, and characterization of the optical isomers of $Co(en)_3^{3+}$ are the objects of this experiment. The preparation of the complex is similar to that used in the preparation of $Co(NH_3)_5Cl^{2+}$ in Experiment 13. A solution of Co^{II} is oxidized in the presence of en to give the Co^{III} complex $Co(en)_3^{3+}$. The oxidant is hydrogen peroxide.

$$CoCl_2 \cdot 6\,H_2O\ +\ 3\ en \cdot 2\,HCl\ \longrightarrow\ [Co(en)_3]Cl_2\ +\ 6\,H_2O\ +\ 6\,HCl$$

$$[Co(en)_3]Cl_2\ +\ \tfrac{1}{2}\,NaOH\ +\ \tfrac{3}{2}\,HCl\ +\ \tfrac{1}{2}\,H_2O_2\ +\ \tfrac{3}{2}\,H_2O$$

$$\longrightarrow\ [Co(en)_3]Cl_3 \cdot \tfrac{1}{2}\,NaCl \cdot 3H_2O$$

In the present experiment, the resulting $[Co(en)_3]Cl_3 \cdot \tfrac{1}{2}\,NaCl \cdot 3\,H_2O$ is isolated as a mixture of the two optically active isomers, and is then resolved by crystallization in the presence of the optically active anion $(+)$tartrate [abbreviated $(+)$tart].

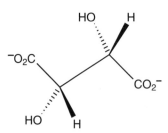

In the present case, $[(+)Co(en)_3][(+)tart]Cl$ is less soluble than $[(-)Co(en)_3][(+)tart]Cl$ and preferentially crystallizes from solution as the pentahydrate:

$$\left.\begin{array}{c} (+)Co(en)_3{}^{3+} \\ (-)Co(en)_3{}^{3+} \end{array}\right\} + (+)tart^{2-} \xrightarrow[H_2O]{Cl^-} \left\{\begin{array}{c} [(+)Co(en)_3][(+)tart]Cl \cdot 5\ H_2O \downarrow \\ [(-)Co(en)_3][(+)tart]Cl \end{array}\right.$$

where the $\downarrow$ symbol next to a formula indicates that the compund precipitates from solution. The $[(+)Co(en)_3][(+)tart]Cl$, whose $[\alpha]_{589.3}$ is $+102°$, is subsequently converted to $[(+)Co(en)_3]I_3 \cdot H_2O$ by addition of I^-. The $[\alpha]_{589.3}$ of the iodide salt is $+89°$.

$$[(+)Co(en)_3][(+)tart]Cl + 3\ I^- \longrightarrow [(+)Co(en)_3]I_3 \cdot H_2O \downarrow + (+)tart^{2-} + Cl^-$$

The other optical isomer, $[(-)Co(en)_3]I_3$, is obtained by adding I^- to the solution from which $[(+)Co(en)_3][(+)tart]Cl \cdot 5\ H_2O$ was previously precipitated. The solid that precipitates with I^- is a mixture of both $(+)$ and $(-)$ isomers of $[Co(en)_3]I_3 \cdot H_2O$, although it is enriched in the $(-)$ isomer. Interestingly, $[(-)Co(en)_3]I_3 \cdot H_2O$ is much more soluble in warm water than the racemate and may be extracted into solution, which on cooling reprecipitates the desired enantiomer, $[(-)Co(en)_3]I_3 \cdot H_2O$, whose $[\alpha]_{589.3} = -89°$. You will establish the optical purities of the isolated $(+)$ and $(-)$ enantiomers by measuring their specific rotations.

Finally, it will be shown that the resolved compound may be racemized by boiling an aqueous solution of one of the enantiomers in the presence of activated charcoal.

EXPERIMENTAL PROCEDURE

Tris(ethylenediamine)cobalt(III) Chloride, [Co(en)$_3$]Cl$_3 \cdot \frac{1}{2}$ NaCl·3 H$_2$O

Put 6.0 g (25 mmol) of $CoCl_2 \cdot 6\ H_2O$, 13.3 g (100 mmol) of ethylenediamine dihydrochloride, and a stir bar in a 250-mL beaker and add approximately 25 mL of water. Stir for a minute or so to dissolve the cobalt salt. The mixture will appear a cloudy pink. Then add 8.0 g (200 mmol) of sodium hydroxide pellets and stir. A cloudy orange solution is formed. Stir for a few minutes until the sodium hydroxide is completely dissolved. Add 20 mL (20 mmol) of 3% H_2O_2 (adjust amounts if a different concentration of hydrogen peroxide is used). The solution darkens upon addition of the peroxide. Dilute the mixture to 50 mL, and heat to boiling for a few minutes on a stirring hot plate until the cloudiness disappears. Remove the stir bar from the hot solution with a "magnet wand", and then place the beaker in ice and let it cool for 30 min. With a suction filtration apparatus (see Fig. 13-1) and a 150-mL medium porosity glass frit, collect the fine orange to yellow-orange needles that have formed in the cooled beaker. Wash the crystals on the frit first with 50 mL of 95% ethanol and then with 20 mL of diethyl ether. Pull air through the crystals on the frit until they are dry and have warmed to room temperature. Record the yield. The filtered solution can be discarded.

Resolution of the Tris(ethylenediamine)cobalt(III) Ion

Put 6.0 g (14.0 mmol) of [Co(en)$_3$]Cl$_3 \cdot \frac{1}{2}$NaCl$\cdot$3H$_2$O, 2.6 g (17.4 mmol) of (+)tartaric acid, and a stir bar in a 100-mL beaker and add 20 mL of water (adjust the quantities if you use a different amount of the cobalt starting material). Then add 1.4 g (35 mmol) of sodium hydroxide, and cover the beaker with a watch glass. Gently stir the mixture, and heat on a stirring hot plate for a few minutes until the solids completely dissolve. Remove the stir bar with a "magnet wand", and then let the solution cool to room temperature overnight. (If the solution is to be left for more than 24 h, seal the mouth of the beaker with Parafilm®.) *Here is a good stopping point for the first day's work.*

After 24 h, collect the parallelepiped-shaped dark orange crystals with a suction filtration apparatus and a 60-mL medium porosity glass frit. Transfer the filtrate to a beaker and save this solution, which contains the other isomer. After you have transferred the filtrate out of the suction filtration apparatus, wash the crystals on the frit first with 20 mL of a 1:1 mixture of water and acetone, and then with 20 mL of pure acetone. Pull air through the crystals of [(+)Co(en)$_3$][(+)tart]Cl$\cdot$5 H$_2$O until they are dry and have warmed to room temperature. Record the yield.

To determine the specific rotation of this compound, dilute approximately 0.5 g of the sample to a solution volume of 10 mL. Introduce this solution into a 1-dm (where dm = decimeter) polarimeter tube, tilting the tube as necessary to remove bubbles from the light path. Following the manufacturer's instructions, first adjust the polarimeter so that the angle of rotation is zero when there is no sample in the instrument. Then introduce the polarimeter tube containing the sample solution into the polarimeter, and measure the sign and magnitude of the rotation. If the instrument requires visual matching of fields, darkening the room is helpful. Calculate $[\alpha]_{589.3}$ for [(+)Co(en)$_3$][(+)tart]Cl $\cdot$ 5 H$_2$O.

To convert the tartrate salt to [(+)Co(en)$_3$]I$_3 \cdot$H$_2$O, put 2.0 g (3.9 mmol) of [(+)Co(en)$_3$][(+)tart]Cl$\cdot$5 H$_2$O in a 50-mL beaker. If the crystals are large, break them into smaller pieces with a spatula. Add one pellet of sodium hydroxide, a stir bar, and 15 mL of water (adjust amounts if more or less than 2.0 g of [(+)Co(en)$_3$][(+)tart]Cl$\cdot$5 H$_2$O is used). With stirring, heat on a stirring hot plate for a few minutes until all the solids dissolve (do not heat for more than 5 min or the cobalt complex will racemize). Add 3.6 g (24 mmol) of NaI. Continue heating and stirring for 1 min, cool the solution in an ice bath, suction filter, and wash the crystals with an ice cold solution of 3 g of NaI in 10 mL of water to remove the tartrate. Wash the crystals with 10 mL of ethanol and then with 10 mL of acetone, allow the [(+)Co(en)$_3$]I$_3 \cdot$H$_2$O to air-dry, and determine the yield. Measure the $[\alpha]_{589.3}$ of the product using a solution of approximately 0.5 g of sample in 10 mL of water.

To isolate [(−)Co(en)$_3$]I$_3 \cdot$H$_2$O, dilute the filtrate from which [(+)Co(en)$_3$][(+)tart]Cl$\cdot$5 H$_2$O was precipitated (see previous discussion) to 30 mL. Add one pellet of sodium hydroxide. Heat the solution until the sodium hydroxide dissolves and add with stirring 8.5 g (57 mmol) of NaI. Cool the solution in an ice bath, collect the impure [(−)Co(en)$_3$]I$_3 \cdot$H$_2$O precipitate by filtration, and wash the precipitate with a solution of 3 g of NaI dissolved in 10 mL of water. To purify, dissolve the precipitate, with stirring, in 35 mL of water at 50 °C. Filter off the undissolved racemate and add 5 g of NaI to the filtrate.

Crystallization of [(−)Co(en)$_3$]I$_3$·H$_2$O occurs on cooling. Collect the precipitate by filtration, wash with 95% ethanol and then with acetone, and finally air-dry. Determine the yield and evaluate $[\alpha]_{589.3}$.

Racemization of (+)Co(en)$_3^{3+}$ or (−)Co(en)$_3^{3+}$

Dissolve approximately 1 g of either [(+)Co(en)$_3$]I$_3$·H$_2$O or [(−)Co(en)$_3$]I$_3$·H$_2$O in a minimum volume of warm water. Add a small amount of activated charcoal and boil the solution for approximately 30 min. Then filter the solution while hot, and add a few grams of NaI to aid in the precipitation of the racemate. Wash the product with a few milliliters each of 95% ethanol and acetone, and then air-dry. Determine $[\alpha]_{589.3}$ of the preciptiated [Co(en)$_3$]I$_3$·H$_2$O racemate.

REPORT

Include the following:
1. Percentage yields of [Co(en)$_3$]Cl$_3$·$\frac{1}{2}$ NaCl·3 H$_2$O, [(+)Co(en)$_3$][(+)tart]Cl· 5 H$_2$O, [(+)Co(en)$_3$]I$_3$·H$_2$O, and [(−)Co(en)$_3$]I$_3$·H$_2$O.
2. The $[\alpha]_{589.3}$ and $[M]_{589.3}$ values for the above complexes. (If an ORD spectrometer is available, record their ORD curves.)
3. If $[\alpha]_{589.3}$ of pure [(+)Co(en)$_3$]I$_3$·H$_2$O is +89°, what percentage of your sample of this compound is actually this enantiomer? Assume that the only impurity is [(−)Co(en)$_3$]I$_3$·H$_2$O. Do the same calculation for your sample of [(−)Co(en)$_3$]I$_3$·H$_2$O.
4. The $[\alpha]_{589.3}$ of the [Co(en)$_3$]I$_3$·H$_2$O, which is isolated from boiling a solution of (+) or (−) Co(en)$_3^{3+}$ with activated charcoal.

PROBLEMS

1. Plot an ORD curve for (−)Co(en)$_3^{3+}$ analogous to that given in Figure 14-3 for (+)Co(en)$_3^{3+}$.
2. If you were to resolve an unknown complex, M(en)$_3^{3+}$, how would you know whether or not your resolved products were optically pure?
3. Draw structures of the geometrical and optical isomers of Co(gly)$_3$, where gly = NH$_2$CH$_2$COO$^-$.
4. Why is it not possible to resolve Co(en)$_3^{2+}$?
5. Draw structures of the optical isomers of Co(EDTA)$^-$, where EDTA = ($^-$O$_2$CCH$_2$)$_2$NCH$_2$CH$_2$N(CH$_2$CO$_2^-$)$_2$. Assume that both nitrogen atoms and one oxygen in each carboxylate "arm" coordinate to Co.
6. In the preparation of Co(en)$_3^{3+}$, what is the purpose of the hydrogen peroxide?
7. In the purification of both (+) and (−) [Co(en)$_3$]I$_3$·H$_2$O, the compounds were washed with water containing NaI. What was the purpose of the NaI?
8. Outline methods for analyzing [(+)Co(en)$_3$]I$_3$·H$_2$O for its Co and I content.
9. The optically active tris(oxalato) complex, Cr(C$_2$O$_4$)$_3^{3-}$, racemizes faster than it exchanges oxalate with free C$_2$O$_4^{2-}$ in solution. This latter fact was determined by using radioactive ^{14}C labeled C$_2$O$_4^{2-}$ in solution. Postulate a mechanism for the racemization.

INDEPENDENT STUDIES

A. Confirm the four-ion nature of $[Co(en)_3]I_3 \cdot H_2O$ by measuring its molar conductance (see Experiment 13).

B. Measure the ^1H NMR spectrum of $[Co(en)_3]I_3 \cdot H_2O$ in D_2O solvent. (Beattie, J. K. *Acc. Chem. Res.* **1971**, *4*, 253.)

C. Prepare and resolve $[Ni(o\text{-phen})_3](ClO_4)_2$, where *o*-phen is 1,10-phenanthroline, into its *d* and *l* enantiomers. (Kauffman, G. B.; Takahashi, L. T. *Inorg. Synth.* **1966**, 8, 227.)

D. Record and compare the UV–vis spectra of the optical isomers of $[Co(en)_3]I_3 \cdot H_2O$.

E. Prepare the cage complexes $[Co(dinosar)]^{3+}$ or $[Co(sepulchrate)]^{3+}$ from racemic or from enantiomerically pure samples of $Co(en)_3^{3+}$. Dinosar and sepulchrate are hexamine ligands that encapsulate the cobalt atom. (Harrowfield, J. M.; Lawrance, G. A.; Sargeson, A. M. *J. Chem. Educ.* **1985**, *62*, 804; Gahan, L. R.; Healy, P. C.; Patch, G. J. *J. Chem. Educ.* **1989**, *66*, 445.)

F. Prepare and resolve the carbon-free optically active compound $[Co\{(OH)_2Co(NH_3)_4\}_3]^{6+}$. (Kauffman, G. B.; Pinnell, R. P. *Inorg. Synth.* **1960**, *6*, 176. Yasui, T.; Ama, T.; Kauffman, G. B. *Inorg. Synth.* **1992**, *29*, 169.)

REFERENCES

Resolution and Characterization of Co(en)$_3$$^{3+}$

Bell, C. F. *Syntheses and Physical Studies of Inorganic Compounds*, Pergamon: New York, 1972, Chapter 22. A summary of physical studies of Co(en)$_3$$^{3+}$.

Broomhead, J. A.; Dwyer, F. P.; Hogarth, J. W. *Inorg. Synth.* **1960**, *6*, 183.

Cramer, R. E.; Huneke, J. T. *Inorg. Chem.* **1975**, *14*, 2565. Infrared spectrum of Co(en)$_3$$^{3+}$.

Techniques of Optical Activity

Lever, A. B. P. *Inorganic Electronic Spectroscopy*, 2nd ed., Elsevier: New York, 1984.

Woldbye, F. in *Technique of Inorganic Chemistry*, Jonassen, H. B.; Weissberger, A., Eds., Wiley: New York, 1965, Vol. IV, p 249. Instrumental methods, theory and survey of the literature of optical rotatory dispersion and circular dichroism.

Optical Activity

Douglas, B. E.; Saito, Y., Eds. *Stereochemistry of Optically Active Transition Metal Compounds*, ACS Symposium Series 119, American Chemical Society: Washington DC, 1980. A compilation of research results in the field.

Hawkins, C. J. *Absolute Configuration of Metal Complexes*, Wiley: New York, 1971. Theory of optical activity in transition metal complexes.

Kauffman, G. B. *Coord. Chem. Rev.* **1974**, *12*, 105. Historical account of Alfred Werner's research on optically active coordination compounds.

Mason, S. F. *Molecular Optical Activity and the Chiral Discriminations*, Cambridge University: New York, 1982. Theory of optical activity and spectroscopic methods.

Saito, Y. *Inorganic Molecular Dissymmetry*, Springer-Verlag: Berlin, 1979. Text on optically active inorganic complexes.

Shields, T. P.; Barton, J. K. *Biochemistry* **1995**, *34*, 15037. The use of optically active Rh complexes to recognize and cleave sequences of DNA.

von Zelewsky, A. *Stereochemistry of Coordination Compounds*, Wiley: New York, 1996.

Wilkins, R. G. *Kinetics and Mechanism of Reactions of Transition Metal Complexes*, 2nd ed., VCH: New York, 1991. Review of rates and mechanisms of coordination compounds.

Metal Dithiolenes and the Use
of Quaternary Ammonium Salts

Note: This experiment requires 7 hours divided between two laboratory periods.

Metal complexes of planar π-conjugated ligands often exhibit rich redox properties because the effects of oxidation or reduction are distributed over many atoms. The ability of π-delocalized complexes to undergo facile oxidation and reduction means that such complexes play important roles in many chemical processes. Porphyrins (Experiment 23) and corrins (Experiment 21) are examples of planar π-conjugated ligands in nature. In hemoglobin, a protein with an iron porphyrin core, the iron center is able to bind oxygen reversibly in part because the effects of the redox process are distributed over the entire Fe(porphyrinate) subunit. Similarly, in the cobalamins the ease of formation and breaking cobalt–carbon bonds is due to the low barriers separating the three oxidation states of cobalt (Experiment 21). In all of these cases, the biological function cannot be mimicked using saturated ligands such as amines. In this experiment, you will study complexes of a family of π-delocalized ligands, the dithiolenes, which have the formula $S_2C_2R_2^{2-}$. Dithiolene ligands are found in all tungsten- and virtually all molybdenum-containing enzymes. These enzymes are involved in redox catalysis, and it is likely that the special π-delocalized character of the dithiolene ligands is critical to their biological function.

Dithiolene ligands form five-membered chelate rings and bind strongly to metals. Two of the best studied families of metal dithiolenes have the formulas $Mo(S_2C_2R_2)_3^{n-}$ and $Ni(S_2C_2R_2)_2^{n-}$, where n can be 0, 1, or 2. These complexes adopt trigonal prismatic and square planar structures, respectively:

When R is an electronegative group such as CN, these complexes are usually isolated as their dianions ($n = 2$). When R is alkyl or aryl, which are relatively electron releasing, then the complexes are more stable as their neutral derivatives ($n = 0$). In all cases, however, these complexes can exist in more than one oxidation state: The dianions can be reversibly oxidized to the monoanions, and the neutral complexes can be reversibly reduced to the monoanions.

One of the most studied dithiolene ligands, and one of the easiest to prepare, has the formula $C_3S_5^{2-}$. This anion is generated by the reduction of CS_2 with alkali metals according to the equation:

The reduction is very slow in the absence of a strongly coordinating solvent. In the present experiment, the reduction is promoted by the slow addition of N,N-dimethylformamide, $(CH_3)_2NCHO$ or DMF. As can be seen from the equation, the reaction generates two products, the desired anion $C_3S_5^{2-}$ and also the trithiocarbonate anion CS_3^{2-}. These species are separated from one another by selective precipitation of the zinc complex $Zn(C_3S_5)_2^{2-}$ as its tetraethylammonium salt:

$$2 \, Na_2C_3S_5 \; + \; ZnSO_4 \; + \; 2 \, N(C_2H_5)_4Br$$

$$\longrightarrow \; [N(C_2H_5)_4]_2[Zn(C_3S_5)_2] \; + \; Na_2SO_4 \; + \; 2 \, NaBr$$

The use of tetraethylammonium bromide to isolate $Zn(C_3S_5)_2^{2-}$ illustrates the utility of quaternary ammonium, or "quat" salts, in inorganic synthesis. Quat salts of large anions are usually insoluble in water but are often highly soluble in polar organic solvents such as acetonitrile, acetone, and dichloromethane. In addition, quat salts are useful in inorganic synthesis because their salts are often nicely crystalline. Similar advantages are afforded by other large organic cations such as $(C_6H_5)_4P^+$, $(C_6H_5)_4As^+$, and $(C_6H_5)_3PNP(C_6H_5)_3^+$ (often abbreviated as PPN^+). Examples of metal-containing anions that are best isolated as salts with large organic cations include $[N(C_4H_9)_4]_2Mo_2O_7$, $(PPN)HFe_3(CO)_{11}$ (see Experiment 18), and $[N(C_2H_5)_4]_3Ti(CN)_6$.

Complementary to the utility of large cations is the use of large anions. For example, many cationic metal complexes have been purified by crystallizing them as salts of large anions such as $CF_3SO_3^-$ (triflate), PF_6^- (hexafluorophosphate), BF_4^- (tetrafluoroborate), and $B(C_6H_5)_4^-$ (tetraphenylborate). In recent years, these anions have increasingly been used instead of ClO_4^- (perchlorate), whose salts can be dangerously explosive. A new generation of alkene polymerization catalysts has been developed that is based on the use of noncoordinating anions such as $B(C_6F_5)_4^-$ and $B[C_6H_3\text{-}3,5\text{-}(CF_3)_2]_4^-$.

Ordinary salts such as NaCl are soluble in water because water forms strong interactions with Na^+ and Cl^-, which are both small ions. Indeed, as you will observe in this experiment, both $Na_2[Zn(C_3S_5)_2]$ and $N(C_2H_5)_4Br$ are highly soluble in water. This solubility is attributable to the strong interactions between the water and Na^+ and Br^- ions, respectively. In contrast, $[N(C_2H_5)_4]_2[Zn(C_3S_5)_2]$ is virtually insoluble in water. You will take advantage of this fact by first precipitating $[N(C_2H_5)_4]_2[Zn(C_3S_5)_2]$ from aqueous solution and later extracting this quat salt into acetone.

Solutions of $Zn(C_3S_5)_2{}^{2-}$ react with benzoyl chloride to give the thioester $C_3S_5(COC_6H_5)_2$, which can be used to prepare other $C_3S_5{}^{2-}$ complexes. This neutral organic product is easily separated from the salt $[N(C_2H_5)_4]_2ZnCl_4$ by virtue of their differing solubilities. In a subsequent step, the thioester is cleaved with sodium methoxide to give a solution of $Na_2C_3S_5$ together with methylbenzoate. Nickel chloride and $Na_2C_3S_5$ combine to give the square planar complex $Ni(C_3S_5)_2{}^{2-}$, which you will isolate from aqueous solution by the addition of the quat salt tetrabutylammonium bromide:

$$[N(C_2H_5)_4]_2[Zn(C_3S_5)_2] \; + \; 4\,C_6H_5COCl \longrightarrow$$

$$2\,C_3S_5(COC_6H_5)_2 \; + \; [N(C_2H_5)_4]_2[ZnCl_4]$$

$$C_3S_5(COC_6H_5)_2 \; + \; 2\,NaOCH_3 \longrightarrow Na_2C_3S_5 \; + \; 2\,C_6H_5CO_2CH_3$$

$$2\,Na_2C_3S_5 \; + \; Ni^{2+} \; + \; 2\,N(C_4H_9)_4{}^+ \longrightarrow [N(C_4H_9)_4]_2[Ni(C_3S_5)_2] \; + \; 4\,Na^+$$

EXPERIMENTAL PROCEDURE

Safety note: Because CS_2 is toxic and has an unpleasant odor, it should be handled in a hood. Sodium metal reacts explosively with water and should be handled with care. Quaternary ammonium salts are toxic. Wear gloves and work in a well-ventilated hood when using them.

The dimethylformamide (DMF) should be dried over Linde 4A molecular sieves before use.

Reduction of CS₂ by Sodium

In a hood, place a 100-mL three-neck round-bottom flask in a small oil bath placed on top of a heater/stirrer (you may use a heating mantle in place of the oil bath but stirring the solution will be more difficult). Equip the flask with a magnetic stir bar, a pressure equalizing addition funnel, a thermometer, and a water-cooled condenser topped with with a gas inlet adapter that is connected to a bubbler and a N_2 source (see Fig. 15-1). Lubricate the joints with silicone grease. Remove the stopper on the addition funnel and flush N_2 through the entire apparatus. Replace the stopper and turn down the nitrogen flow so that a bubble appears every few seconds. Weigh out between 1.05 and 1.25 g (~ 0.05 mol) of sodium; use a paper towel to wipe off most of the oil from the sodium. Cut the weighed sodium into about eight small pieces.

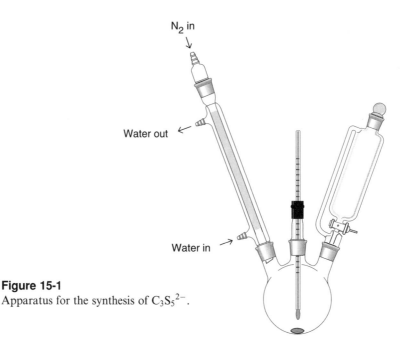

N₂ in

Water out ←

Water in →

Figure 15-1
Apparatus for the synthesis of $C_3S_5^{2-}$.

Remove the thermometer, degrease the joint, and add the sodium and 10 mL of carbon disulfide to the flask. Replace the thermometer in the joint and turn on the stirrer. Into the addition funnel (make sure that the stopcock on the funnel is closed), place 10 mL of DMF that has previously been dried over Linde 4A molecular sieves. Place a safety shield between you and the apparatus. Add about 0.5 mL of the DMF to the CS_2/Na slurry (estimate the amount; it is not necessary to be highly accurate). Turn the heater on and heat the slurry to about 50 °C. Once the reaction starts (as judged by the appearance of a red color), turn off the heat and add the DMF dropwise in order to keep the CS_2 boiling. The addition of the DMF should take about 30 min. After the DMF has been added, turn the heater back on and keep the solution at reflux. While the reaction mixture is being refluxed, prepare the following two solutions in 50-mL Erlenmeyer flasks:

Zn²⁺ solution: Dissolve 1.9 g (0.0075 mol) of $ZnSO_4·7H_2O$ in 20 mL of concentrated NH_4OH.
Quat salt solution: Dissolve 2.6 g (0.0125 mol) of $N(C_2H_5)_4Br$ in 12 mL of H_2O.

$[N(C_2H_5)_4]_2[Zn(C_3S_5)_2]$
After the sodium/CS_2 reaction mixture has been refluxed for 1.5 h, its color should be deep red to black. Cool the reaction mixture to room temperature. Making sure that the stopcock on the dropping funnel is closed, charge the dropping funnel with 20 mL of methanol. While stirring the cooled reaction mixture, add the methanol in about 1-mL portions every 30 seconds or so. Stir the mixture until any remaining pieces of sodium are destroyed. Now remove the

dropping funnel, thermometer, and condenser, and pour the reaction mixture into a 250-mL Erlenmeyer flask. Rinse the 100-mL reaction flask with 50 mL of H_2O and add the rinse to the Erlenmeyer flask. Add the zinc solution to the stirred reaction mixture. A few minutes later, add the quat salt solution and cool the mixture in ice for 10 min. Collect the resulting bright red precipitate by suction filtration (see Fig. 13-1) and wash it with 50 mL of H_2O and 10 mL of methanol. Dispose of the filtrate in a hood by first treating it with ~ 1 M aqueous NaOH, which destroys the CS_2. Dry the product in air and, if time permits, determine its melting point. *Here is a good stopping point for the first day's work.*

$C_3S_5(COC_6H_5)_2$

Safety note: Benzoyl chloride is corrosive, toxic, and a lachrimator (tear gas). It reacts violently with water with production of corrosive fumes. Use it in a hood.

In a hood, dissolve 2.0 g (2.8 mmol) of $[N(C_2H_5)_4]_2[Zn(C_3S_5)_2]$ in 50 mL of acetone and then add 1.6 mL (1.6 g, 11 mmol) of benzoyl chloride (*not* benzyl chloride!). A yellow precipitate should form. Allow the reaction mixture to stand for a few minutes, and then cool it in ice for 10 min. Collect the yellow-orange solid by suction filtration. Dissolve the solid in 100 mL of dichloromethane, add an equal volume of methanol, and concentrate this solution to about 40 mL on a rotary evaporator (while the solution is concentrating, you may wish to prepare the $NiCl_2$ and $N(C_4H_9)_4Br$ solutions needed to synthesize the nickel complex in the next section). Collect the fine yellow needles of $C_3S_5(COC_6H_5)_2$ by filtration and dry them in air. Record the yield. Although your product will be used in the next step, save some of it so that you can measure its infrared (IR) spectrum.

$[N(C_4H_9)_4]_2[Ni(C_3S_5)_2]$

Dissolve 0.18 g (0.75 mmol) of nickel(II) chloride hexahydrate, $NiCl_2 \cdot 6\ H_2O$, in 10 mL of methanol. Dissolve 0.5 g (1.7 mmol) of tetrabutylammonium bromide, $N(C_4H_9)_4Br$, in 5 mL of methanol. Prepare a solution of sodium methoxide, $NaOCH_3$, by dissolving 0.24 g (6 mmol) of NaOH pellets in ~ 2 mL of water in a 50 mL Erlenmeyer flask (use magnetic stirring), and then adding 15 mL of methanol. Add 0.7 g (1.7 mmol) of $C_3S_5(COC_6H_5)_2$. Warm the mixture on a hot plate and stir the resulting bright red solution for 10 min. To this solution add the Ni^{2+} solution. Stir the solution for a few minutes, and then add the $N(C_4H_9)_4^+$ solution. After a few minutes, filter off the dark green precipitate. Rinse the solid with 10 mL of methanol. Record the yield and characterize your product by IR and ultraviolet–visible (UV–vis) spectroscopy.

REPORT

Include the following:
1. Yields of all products.
2. IR spectra of all products.
3. ^{1}H NMR spectrum of $C_3S_5(COC_6H_5)_2$ and assignment of the peaks.
4. UV–vis spectrum of $[N(C_4H_9)_4]_2[Ni(C_3S_5)_2]$.

PROBLEMS

1. The first step in this synthesis is the reduction of CS_2 to CS_2^-. What known triatomic molecule has the same electron count as CS_2^-, and what is its geometry?
2. List the advantages of using tetraalkylammonium cations to isolate anionic transition metal complexes.
3. The stability of $N(C_4H_9)_4^+$ salts often depends on the basicity of the anion. For example, anhydrous $N(C_4H_9)_4F$ is not known, whereas $N(C_4H_9)_4Cl$ is stable. What reaction occurs when a strong base reacts with a tetraalkylammonium cation?

INDEPENDENT STUDIES

A. Oxidize $Ni(C_3S_5)_2^{2-}$ to the monoanion and characterize it by electron paramagnetic resonance (EPR) spectroscopy. (Steimecke, G.; Sieler, H. J.; Kirmse, R.; Hoyer, E. *Phosphorus Sulfur* **1979**, *7*, 49.)
B. Treat the $C_3S_5^{2-}$ solution with $(C_5H_5)_2TiCl_2$ to give $(C_5H_5)_2TiC_3S_5$. (Yang, X.; Rauchfuss, T. B.; Wilson, S. R. *J. Am. Chem. Soc.* **1988**, *111*, 3465.)
C. Measure the cyclic voltammogram of $[N(C_4H_9)_4]_2[Ni(C_3S_5)_2]$. (Valade, L.; Cassoux, P.; Gleizes, A.; Interrante, L. V. *J. Phys. (Paris)* **1983**, *44*, C3, 1183.)
D. Examine the effect of solvent polarity on the optical spectrum of $[N(C_4H_9)_4]_2[Ni(C_3S_5)_2]$.

REFERENCES

Dithiolene Complexes

Cassoux, P.; Valade, L.; Kobayashi, H.; Kobayashi, A.; Clark, R. A.; Underhill, A. E. *Coord. Chem. Rev.* **1991**, *110*, 115. Synthesis of electrical conductors from metal dithiolenes.

Collison, D.; Garner, C. D. Joule, J. A. *Chem. Soc. Rev.* **1996**, 25. A review of the chemistry and biochemistry of molybdenum dithiolene complexes.

Kim, M. K.; Mukund, S.; Kletzin, A.; Adams, M. W. W.; Rees, D. C. *Science* **1995**, *267*, 1467. The structure of a tungsten-containing enzyme, consisting of two dithiolene ligands bound to the tungsten atom.

Mueller-Westerhoff, U. T.; Vance, B. in *Comprehensive Coordination Chemistry*, Wilkinson, G.; Gillard, R. D.; McCleverty, J. A., Eds., Pergamon: New York, 1987; Vol. 2., p 595. Review of dithiolene complexes.

Olk, R.-M.; Olk, B.; Dietzsch, W.; Kirmse, R.; Hoyer, E. *Coord. Chem. Rev.* **1992**, *117*, 99. Survey of $C_3S_5^{2-}$ complexes.

Ramao, M. J.; Archer, M.; Moura, I.; Moura, J. J. G.; LeGall, J.; Engh, R.; Schneider, M.; Hof, P.; Huber, R. *Science* **1995**, *270*, 1170. Structure of a molybdenum dithiolene enzyme.

Steimecke, G.; Sieler, H. J.; Kirmse, R.; Hoyer, E. *Phosphorus Sulfur* **1979**, 7, 49. First synthesis of complexes of $C_3S_5^{2-}$.

Yu, L.; Zhu, D. *Phosphorus, Sulfur, Silicon* **1996**, *116*, 225. Rapid synthesis of the $C_3S_5^{2-}$ anion.

Specialized Anions and Cations in Inorganic Synthesis

Brookhart, M.; Grant, B.; Volpe, Jr., A. F. *Organometallics* **1992**, *11*, 3920. Synthesis of the $B[C_6H_3\text{-}3,5\text{-}(CF_3)_2]_4{}^-$ anion.

Christie, K. O.; Dixon, D. A.; Mercier, H. P. A.; Sanders, J. C. P.; Schrobilgen, G. J.; Wilson, W. W. *J. Am. Chem. Soc.* **1994**, *116*, 2850. Use of $N(CH_3)_4{}^+$ to isolate unusual fluoro anions such as $PF_4{}^-$.

Entley, W. R.; Wilson, S. R.; Girolami, G. S. *J. Am. Chem. Soc.* **1997**, *119*, 6251. Use of $N(C_2H_5)_4{}^+$ to isolate the hexacyanotitanate ion.

Klemperer, W. G. *Inorg. Synth.* **1990**, *27*, 74. Use of $N(C_4H_9)_4{}^+$ to isolate isopolyoxo-metalates.

Piers, W. E.; Chivers, T. *Chem. Soc. Rev.* **1997**, *26*, 345. Concise review of per-fluorophenylborates as noncoordinating anions.

Strauss, S. H. *Chem. Rev.* **1993**, *93*, 927. A review of weakly coordinating anions.

Wagner, M. J.; Dye, J. L. *Ann. Rev. Mater. Sci.* **1993**, *23*, 223. Use of crown ethers and cryptand ligands in the synthesis of alkali metal anions.

Part IV

ORGANOMETALLIC CHEMISTRY

The Metal–Arene Complex
$[1,3,5\text{-}C_6H_3(CH_3)_3]Mo(CO)_3$

Note: This experiment requires one 3- or 4-hour laboratory period.

The discovery of ferrocene and related metallocenes in the early 1950s led to an explosion of interest in organometallic chemistry that continues to this day. This discovery showed that the π bonds in unsaturated hydrocarbons (such as the cyclopentadienide anion) could form extremely strong links to transition metals (see Experiment 20). Because benzene is similar in many respects to the cyclopentadienide anion (both have six-electron π systems), inorganic chemists became interested in whether benzene could also form complexes with transition metals. In 1955, the first such complex was described: the compound bis(benzene)chromium. As it turned out, compounds closely related to bis(benzene)chromium had actually been prepared as early as 1919, but their structures had been incorrectly drawn.

The standard synthesis of $Cr(C_6H_6)_2$ is carried out according to the following equations:

$$3\,CrCl_3 + 2\,Al + AlCl_3 + 6\,C_6H_6 \longrightarrow 3\,Cr(C_6H_6)_2{}^+ + 3\,AlCl_4{}^-$$

$$2\,Cr(C_6H_6)_2{}^+ + S_2O_4{}^{2-} + 4\,OH^- \longrightarrow 2\,Cr(C_6H_6)_2 + 2\,SO_3{}^{2-} + 2\,H_2O$$

In the first reaction, the $AlCl_3$ acts to remove Cl^- from the Cr^{III} as $AlCl_4{}^-$, and powdered Al reduces the Cr^{III} to Cr^I. The cationic complex $Cr(C_6H_6)_2{}^+$ is then reduced with dithionite ion, $S_2O_4{}^{2-}$, to give $Cr(C_6H_6)_2$ in an overall yield of 60%. Bis(benzene)chromium(0) has a sandwich structure similar to that of ferrocene:

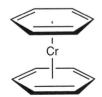

161

It is possible to prepare a variety of other bis(benzene) complexes by reaction of other transition metal halides with Al and $AlCl_3$ in benzene, often followed by reduction with $S_2O_4^{2-}$. Some of these are $V(C_6H_6)_2^+$, $V(C_6H_6)_2$, $Mo(C_6H_6)_2^+$, $Mo(C_6H_6)_2$, $W(C_6H_6)_2^+$, $W(C_6H_6)_2$, $Re(C_6H_6)_2^+$, $Fe(C_6H_6)_2^{2+}$, $Ru(C_6H_6)_2^{2+}$, and $Co(C_6H_6)_2^+$. More recently, many of these compounds have been prepared by a completely different route involving the direct reaction of benzene with metal atoms generated by evaporating the metal from a melted ingot under vacuum.

In some respects, benzene acts like three alkenes and donates six electrons to the metal. As illustrated by the present experiment, benzene and its derivatives do often react to displace three donor ligands. The three ligands that will be displaced in this synthesis are CO groups. Because mesitylene, $1,3,5-C_6H_3(CH_3)_3$, forms more stable complexes than benzene, it will be used as the aromatic ligand.

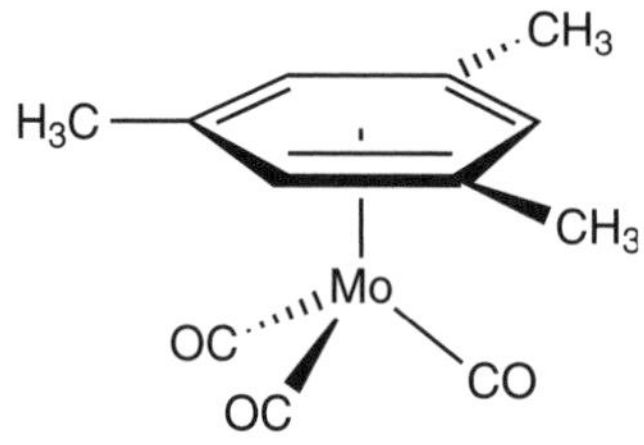

The synthesis of $[1,3,5-C_6H_3(CH_3)_3]Mo(CO)_3$ is carried out by refluxing a solution of $Mo(CO)_6$ in mesitylene.

In $[1,3,5-C_6H_3(CH_3)_3]Mo(CO)_3$, the benzene ring lies parallel to the plane of the three carbon atoms of the CO groups, as shown above. It may be considered to be an octahedral complex because all of the OC–Mo–CO bond angles are very nearly 90° as they are in $Mo(CO)_6$. The arene group occupies one face of the octahedron. Although early researchers thought that benzene might adopt a "frozen Kekulé" structure when coordinated to a metal atom, X-ray structural studies show that all of the C–C bond distances in the benzene ring are equal, so that the π system is still delocalized.

By reaction of other aromatic molecules with $Mo(CO)_6$ (or its first- and third-row analogues $Cr(CO)_6$ and $W(CO)_6$), it is possible to synthesize compounds with unusual geometries:

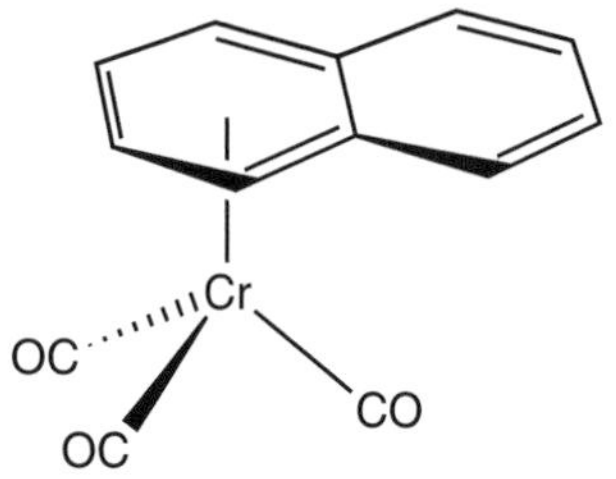

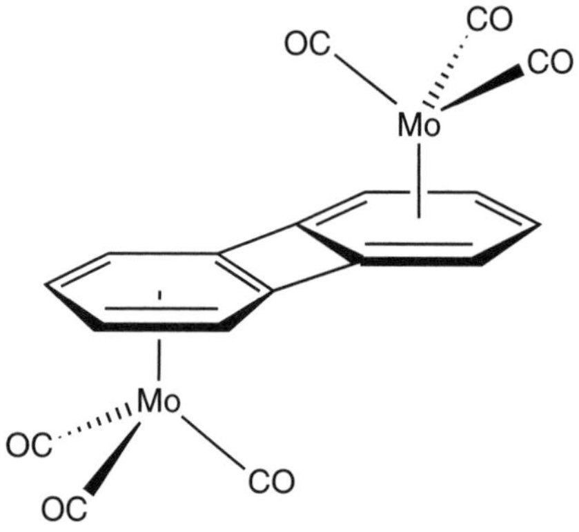

The analogy of borazole, $B_3N_3H_6$ (sometimes called "inorganic benzene"), to benzene led to the synthesis of the hexamethylborazole analogue of [$C_6(CH_3)_6$]Cr(CO)$_3$:

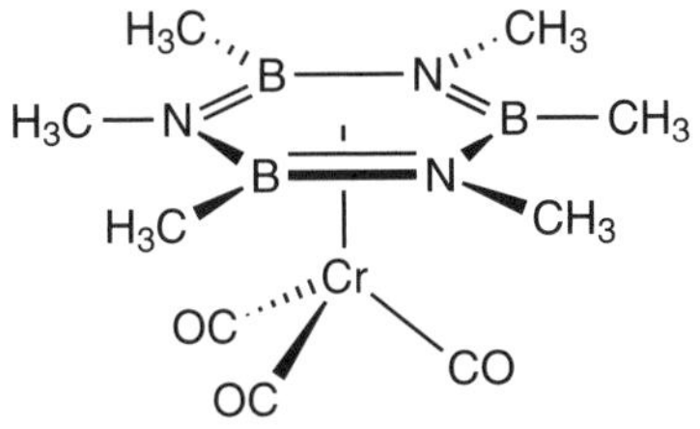

Of some interest to organometallic chemists is the aromaticity of benzene in these complexes compared with that of the free molecule. A chemical indication of the aromaticity of a benzene derivative is its rate of acetylation under Friedel–Crafts reaction conditions: Electrophilic attack of CH_3CO^+ on the ring is faster for benzene rings having more electron density in the π system. The acylation of (C_6H_6)Cr(CO)$_3$ occurs readily but at a rate that is slower than for benzene itself. This difference in rate suggests that the electron density in the benzene ring is depleted when it is complexed to a Cr(CO)$_3$ unit. This result confirms that benzene, like most ligands, acts as an electron donor toward transition metals.

Infrared Spectra of Metal Carbonyls

The infrared (IR) spectra of metal carbonyls contain intense bands near 2000 cm^{-1} due to the C–O stretching vibrations of the metal-bound carbon monoxide ligands. For compounds that contain more than one CO ligand, the number of bands and their relative intensities depend on how the carbonyl ligands are arranged in space relative to one another. Here we will discuss some aspects of how the spectra of metal carbonyls can be analyzed to determine the relative arrangement of CO ligands about a metal center.

For a complex that contains one carbonyl ligand, such as IrCl(CO)[P(C_6H_5)$_3$]$_2$ described in Experiment 19, only one carbon–oxygen stretching band is seen. When two or more carbonyl ligands are attached to a metal center, they vibrate in concert—that is, the C–O bonds stretch and contract in synchrony. For a metal complex with two carbonyl ligands, M(CO)$_2$L$_x$ (where L$_x$ represents all the other ligands attached to the metal), there are two possibilities: (1) when one C–O bond stretches, the other one does too, and (2) when one C–O bond stretches, the other one contracts. Each of these possibilities is called a "stretching mode": The former is called the symmetric stretch, and the latter is called the antisymmetric stretch. The two modes are shown schematically in the following diagram:

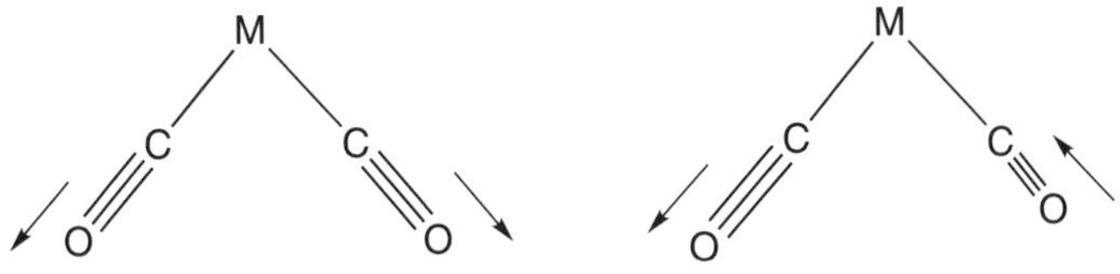

$(\nu_{CO})_{sym}$ $(\nu_{CO})_{asym}$

The absorption frequencies of the symmetric and antisymmetric stretching vibrations are different (for metal carbonyls, the symmetric stretch is generally higher in frequency), and thus in general the IR spectrum of a metal dicarbonyl complex $M(CO)_2L_x$ will exhibit two C–O stretching bands.

The relative intensity of the two IR bands depends on the C–M–C angle between the two carbonyl ligands. An equation has been derived that gives an approximate relationship between the intensities and the C–M–C angle (which we will call θ):

$$\frac{I_{\text{sym}}}{I_{\text{anti}}} = \tan^2\left(\frac{\theta}{2}\right)$$

If the two carbonyl ligands are cis to one another (so that $\theta \approx 90°$), then $\tan^2(\theta/2) = 1$ and the symmetric and antisymmetric bands will have equal intensities. If the two carbonyl ligands are trans to one another (so that $\theta = 180°$), then $\tan^2(\theta/2) = 0$ and the symmetric stretch is vanishingly weak (has zero intensity). In this case, only one CO stretching band will be seen in the IR spectrum, despite the fact that both the symmetric and antisymmetric vibrations still occur. Only one band is seen because the symmetric stretching vibration is forbidden by symmetry to absorb IR radiation. By using a different form of vibrational spectroscopy, Raman spectroscopy, one can observe the symmetric CO stretching band for a trans dicarbonyl. You can learn more about Raman spectroscopy by consulting a physical chemistry textbook.

If a metal has three carbonyl ligands, as in $[1,3,5\text{-}C_6H_3(CH_3)_3]Mo(CO)_3$, then there will be three C–O stretching modes. As we will see, however, there may be fewer than three C–O stretching bands in the IR spectrum if the three carbonyl ligands are arranged around the central metal in certain special ways. We will only discuss the case that is relevant to the present Experiment: a metal complex in which the three carbonyl ligands are in identical environments and are arranged in a pyramidal fashion. There are three C–O stretching modes for a pyramidal arrangement of three CO ligands. One is called the symmetric stretch, in which all three C–O bonds stretch simultaneously. The other two modes are antisymmetric: As one C–O bond stretches, one or both of the other two C–O bonds contracts:

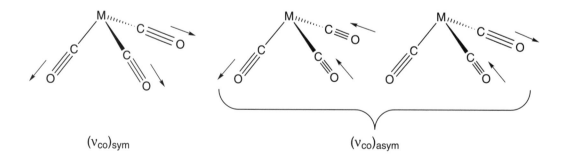

$(\nu_{CO})_{\text{sym}}$ $(\nu_{CO})_{\text{asym}}$

With the help of a mathematical approach called group theory, it can be shown that the two antisymmetric modes have exactly identical absorption frequencies. This means that there will be only two observable bands near 2000 cm^{-1} in the IR spectrum of a pyramidal tricarbonyl complex: one for the symmetric mode and another for the coinciding antisymmetric modes. The relative intensity of these two bands again provides information about the angles between the CO ligands. The following equation gives this relationship:

$$\frac{I_{\text{sym}}}{I_{\text{anti}}} = \frac{3\cot^2(\theta/2) - 1}{4} \qquad (1)$$

where θ is the C–M–C angle between every pair of carbonyl ligands. If the C–M–C angle is 75.5°, then $I_{\text{sym}}/I_{\text{anti}}$ will be equal to 1.0 and the two bands will be equal in intensity. You will use Eq. 1 to analyze the IR spectrum of [1,3,5-C$_6$H$_3$(CH$_3$)$_3$]Mo(CO)$_3$ and to determine the C–M–C angles in this complex.

What about complexes with more than three carbonyl ligands? Interestingly, the IR spectra of octahedral hexacarbonyl complexes such as Mo(CO)$_6$ contain only one C–O stretching band. Obviously, the number of C–O stretching bands seen in an IR spectrum does not directly indicate how many CO ligands are present in a particular complex. If the number of carbonyl ligands can be established by other means, however (such as mass spectrometry), IR spectroscopy can often be used to determine how the carbonyl groups are arranged about the metal center.

EXPERIMENTAL PROCEDURE

Safety note: liquid metal carbonyls, in general, are toxic compounds and should be handled with care. Solid carbonyl complexes such as Mo(CO)$_6$ do not present as serious a danger but they should be handled in a hood. In the following reaction, carbon monoxide gas is evolved. The volume of CO given off is relatively small, however, and if the reaction is conducted in a hood, the CO will not be present in sufficiently high concentrations to be dangerous.

(Mesitylene)tricarbonylmolybdenum(0), [1,3,5-C$_6$H$_3$(CH$_3$)$_3$]Mo(CO)$_3$

The reaction will be conducted in the apparatus shown in Figure 16-1. Assemble the apparatus in the hood. First put 2.0 g (7.6 mmol) of Mo(CO)$_6$ and 10 mL (72 mmol) of mesitylene (bp, 165 °C) in a 50-mL three-neck round-bottom flask with a side-arm and stopcock. In the central joint, place a straight reflux condenser of approximately 30-cm length that is topped with a gas inlet connected to a bubbler (see Fig. I-4). Use silicone grease to lubricate the joint. The Mo(CO)$_6$ is volatile and will sublime into the condenser during the reaction. For this reason, it is desirable for the mesitylene vapors to rise high into the condenser to return any sublimed Mo(CO)$_6$ to the reaction flask. To allow the mesitylene to wash the Mo(CO)$_6$ into the flask, do not cool the condenser with water. The air in the room will provide adequate cooling.

The tendency of Mo(CO)$_6$ and the product to react with oxygen at high temperatures requires that the reaction be conducted in an inert atmosphere. Connect the gas inlet at the top of the condenser to a N$_2$ cylinder via a length of

N₂ in

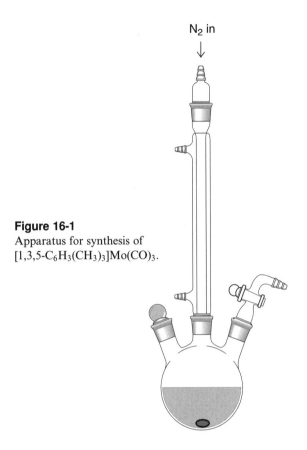

Figure 16-1
Apparatus for synthesis of
$[1,3,5\text{-}C_6H_3(CH_3)_3]Mo(CO)_3$.

rubber tubing. Plug the third joint, and flush the apparatus with a moderate stream of N_2 for approximately 5 min. Turn off the N_2 and then close the stopcock. Heat the solution at a moderate boil for about 30 min with a rheostat-controlled heating mantle. Then remove the heating mantle and immediately turn on the N_2 flush to prevent the bubbler from sucking back.

When the solution has cooled to room temperature, turn off the N_2 stream and dismantle the apparatus. Degrease the joints by wiping them with a tissue. Add 15 mL of hexane (or hydrocarbon fraction of 60–70 °C boiling range) to complete the precipitation. Suction filter (Fig. 13-1) the solution on a medium frit and rinse, with 5 mL of hexane, the yellow product that is contaminated with black metallic molybdenum. Purify the crude product by dissolving it in a minimum of CH_2Cl_2 (about 10 mL). After filtering the solution, add 25 mL of hexane to precipitate the product. Suction filter off the yellow $[1,3,5\text{-}C_6H_3(CH_3)_3]Mo(CO)_3$, wash it twice with 4 mL of hexane, and allow the product to dry on the frit while under aspiration. (A second batch of product can be obtained by reducing the volume of the filtrate in a rotary evaporator.) Calculate the percentage yield. If desired, the product may be further purified by sublimation at approximately 120 °C under high vacuum (Fig. 20-3). Because

[1,3,5-C$_6$H$_3$(CH$_3$)$_3$]Mo(CO)$_3$ will decompose over a period of weeks in the presence of light and air, it should be stored in the dark in a tightly stoppered container that has been flushed with N$_2$.

Solution Infrared Spectrum

Determine the IR spectrum of [1,3,5-C$_6$H$_3$(CH$_3$)$_3$]Mo(CO)$_3$ by dissolving a few small crystals in approximately 2 mL of CH$_2$Cl$_2$ (see Experiment 17 for further details). With a syringe or pipet, introduce the solution into an IR cell such as that in Figure 17-4. If possible, plot the IR spectrum with the y axis showing absorbance, *not* transmittance, so that you can measure the relative intensities of the two bands. The intensity of a band is proportional to how much light is absorbed, not how much is transmitted.

   If the two absorption features had identical widths, then the maximum absorbance could be used as a measure of the intensities of the C–O stretching bands. Owing to solvent effects that broaden one of the features more than the other, it is necessary instead to measure the areas of the two peaks; these numbers will be better measures of the intensities. Follow the procedure given by your instructor to measure the areas of the two peaks. From this information, determine the ratio I_{sym}/I_{anti} and from this ratio calculate the C–M–C angle θ in [1,3,5-C$_6$H$_3$(CH$_3$)$_3$]Mo(CO)$_3$ using Eq. 1.

REPORT

Include the following:

1. Percentage yield of [1,3,5-C$_6$H$_3$(CH$_3$)$_3$]Mo(CO)$_3$ and discussion of factors responsible for making the yield less than 100%.
2. Infrared spectrum and the assignment of absorptions to C–O, C–H, and C–C vibrations.
3. Calculation of the C–M–C angle in [1,3,5-C$_6$H$_3$(CH$_3$)$_3$]Mo(CO)$_3$.

PROBLEMS

1. If the reaction of Mo(CO)$_6$ with mesitylene were conducted in the presence of air, what would probably be the decomposition products?
2. Why does the symmetric C–O stretch of a trans dicarbonyl complex have zero intensity?
3. How would the C–O stretching frequencies vary in the compounds (C$_6$H$_6$)Mo(CO)$_3$, [1,3,5-C$_6$H$_3$(CH$_3$)$_3$]Mo(CO)$_3$, and [C$_6$(CH$_3$)$_6$]Mo(CO)$_3$? Explain your reasoning.
4. The compound [1,3,5-C$_6$H$_3$(CH$_3$)$_3$]Mo(CO)$_3$ reacts with P(OCH$_3$)$_3$ to form *fac*-Mo(CO)$_3$[P(OCH$_3$)$_3$]$_3$ according to the rate law

$$\text{Rate} \ = \ k[(1,3,5\text{-C}_6\text{H}_3(\text{CH}_3)_3)\text{Mo(CO)}_3][\text{P(OCH}_3)_3]$$

   Postulate a mechanism for the reaction. Would you expect the same reaction with PF$_3$ to proceed slower or faster than that observed for P(OCH$_3$)$_3$ and why?

5. How would you determine in an afternoon's experiment whether or not your isolated $[1,3,5\text{-}C_6H_3(CH_3)_3]Mo(CO)_3$ was pure?

6. In the reaction of $Mo(CO)_6$ with mesitylene, which method would give the higher yield of $[1,3,5\text{-}C_6H_3(CH_3)_3]Mo(CO)_3$: (a) as performed in the experiment or (b) in a sealed apparatus such as an autoclave? Explain.

7. If you wished to carry out the following reaction, $(C_6H_6)Mo(CO)_3 + 1,3,5\text{-}C_6H_3(CH_3)_3 \longrightarrow [1,3,5\text{-}C_6H_3(CH_3)_3]Mo(CO)_3 + C_6H_6$, what reaction conditions (solvent, temperature, amounts of reactants, and so on) would you use? Would you expect the equilibrium constant for the reaction to be greater or less than 1?

8. If you wished to carry out the reverse of the reaction performed in this experiment, that is, $[1,3,5\text{-}C_6H_3(CH_3)_3]Mo(CO)_3 + 3\ CO \longrightarrow Mo(CO)_6 + 1,3,5\text{-}C_6H_3(CH_3)_3$, what reaction conditions and techniques would you use?

INDEPENDENT STUDIES

A. Use $[1,3,5\text{-}C_6H_3(CH_3)_3]Mo(CO)_3$ to catalyze the polymerization of phenylacetylene. (Wood, P. S.; Farona, M. F. *J. Polym. Sci., Polym. Chem. Ed.* **1974**, *12*, 1749. Vijayaraj, T. A.; Sundararajan, G. *Organometallics* **1997**, *16*, 4940.)

B. Measure the ^{1}H NMR spectrum of $[1,3,5\text{-}C_6H_3(CH_3)_3]Mo(CO)_3$ in chloroform or deuterochloroform. (Measure its NMR spectrum soon after preparing the sample, because some decomposition will occur on standing.) Compare the chemical shifts you measure with those seen for mesitylene in $CHCl_3$, which has absorptions at δ 2.25 and 6.78 with relative intensities of 9 to 3, respectively.

C. Prepare (arene)$Mo(CO)_3$ complexes of other arenes such as $C_6H_5N(CH_3)_2$ or $C_6(CH_3)_6$. Compare their IR and NMR spectra with those of the mesitylene complex. (Mahaffy, C. A. L.; Pauson, P. L. *Inorg. Synth.* **1990**, *28*, 136. Iverson, D. J.; Hunter, G.; Blount, J. F.; Damewood, J. R.; Mislow, K. *J. Am. Chem. Soc.* **1981**, *103*, 6073.)

D. Prepare and characterize (η^6-cycloheptatriene)$Mo(CO)_3$ and study its reactions. (Timmers, F. J.; Wacholtz, W. F. *J. Chem. Educ.* **1994**, *71*, 987.)

E. Prepare and characterize *fac*-$Mo(CO)_3[P(OCH_3)_3]_3$ obtained from the reaction of $[1,3,5\text{-}C_6H_3(CH_3)_3]Mo(CO)_3$ with $P(OCH_3)_3$. (Pidcock, A.; Smith, J. D.; Taylor, B. W. *J. Chem. Soc. A* **1967**, 872. Zingales, F.; Chiesa, A.; Basolo, F. *J. Am. Chem. Soc.* **1966**, *88*, 2707.)

F. Prepare and characterize $[1,3,5\text{-}C_6H_3(CH_3)_3]Cr(CO)_3$ by treating $Cr(CO)_6$ with a few equivalents of mesitylene in a refluxing mixture of 90% di(*n*-butyl)ether and 10% tetrahydrofuran.

G. Prepare and characterize a salt of $[1,3,5\text{-}C_6H_3(CH_3)_3]_2Fe^{2+}$. (Helling, J. F.; Braitsch, D. M. *J. Am. Chem. Soc.* **1970**, *92*, 7207. Helling, J. F.; Rice, S. L.; Braitsch, D. M.; Mayer, T. *J. Chem. Soc., Chem. Commun.* **1971**, 930.)

H. Prepare and characterize (*p*-cymene)$_2Ru_2Cl_4$. (Bennett, M. A.; Huang, T.-N.; Matheson, T. W.; Smith, A. K. *Inorg. Synth.* **1982**, *21*, 74.)

I. Use group theory to analyze the stretching modes in [1,3,5-C$_6$H$_3$(CH$_3$)$_3$]Mo(CO)$_3$, *trans*-Mo(CO)$_4$[P(C$_6$H$_5$)$_3$]$_2$, and Mo(CO)$_6$. How many C–O stretching bands will be seen for each?

REFERENCES

Metal Arene Complexes

Adams, D. M.; Christopher, R. E.; Stevens, D. C. *Inorg Chem.* **1975**, *14*, 1562. Infrared and Raman studies of (C$_6$H$_6$)Cr(CO)$_3$.

Armstrong, R. S.; Aroney, M. J.; Barnes, C. M.; Nugent, K. W. *Appl. Organomet. Chem.* **1990**, *4*, 569. Infrared and Raman studies of [1,3,5-C$_6$H$_3$(CH$_3$)$_3$]Mo(CO)$_3$.

Byers, B. P.; Hall, M. B. *Inorg. Chem.* **1987**, *26*, 2186. X-ray structural study of [C$_6$(CH$_3$)$_6$]Cr(CO)$_3$.

Brownlee, R. T. C.; O'Connor, M. J.; Shehan, B. P.; Wedd, A. G. *Aust. J. Chem.* **1986**, *39*, 931. The NMR and IR spectra of (arene)Mo(CO)$_3$ compounds.

Fischer, E. O.; Öfele, K.; Essler, H.; Fröhlich, W.; Mortensen, J. P.; Semmlinger, W. *Chem. Ber.* **1958**, *91*, 2763. Preparation of (arene)M(CO)$_3$ compounds.

Kane-Maguire, L. A.; Honig, E. D.; Sweigart, D. A. *Chem. Rev.* **1984**, *84*, 525. A review of arene complexes of the transition metals.

Nicholls, B.; Whiting, M. C. *J. Chem. Soc.* **1959**, 551. Preparation of (arene)M(CO)$_3$ compounds.

Solladié-Cavallo, A. *Polyhedron* **1985**, *4*, 901. A review of the chemistry of (arene)Cr(CO)$_3$.

Timms, P. L. *Chem. Soc. Rev.* **1996**, *93*, 25. Survey of the use of metal vapors in the synthesis of metal arene complexes and other species.

Wang, Y.; Angermund, K.; Goddard, R.; Krüger, C. *J. Am. Chem. Soc.* **1987**, *109*, 587. X-ray structural study of (C$_6$H$_6$)Cr(CO)$_3$.

IR Spectroscopy

Drago, R. S. *Physical Methods for Chemists*, Saunders: Orlando, FL, 1992. Chapters 6 and 7. Basic treatment of IR and NMR spectroscopy, with applications to inorganic compounds.

King, R. B. *Organometallic Syntheses*, Academic: New York, 1965; Vol. 1. Techniques of synthesis and characterization of organometallic compounds, including several arene complexes.

Shriver, D. F.; Drezdzon, M. A. *The Manipulation of Air-sensitive Compounds*, 2nd ed., Wiley: New York, 1986. Excellent coverage of techniques for handling air-sensitive compounds.

Tsutsui, M., Ed. *Characterization of Organometallic Compounds*, Wiley: New York, 1969; Vols. 1 and 2. A very useful general reference to instrumental methods of characterizing organometallic compounds.

Group Theory Applied to IR Spectroscopy

Braterman, P. S. *Metal Carbonyl Spectra*, Academic: New York, 1975. The standard text on the analysis of the vibrational spectra of metal carbonyls.

Cotton, F. A. *Chemical Applications of Group Theory*, 3rd ed., Wiley: New York, 1990.

Cotton, F. A.; Kraihanzel, C. S. *J. Am. Chem. Soc.* **1962**, *84*, 4432. Force constant model for analyzing the IR spectra of metal carbonyls.

General References to Organometallic Chemistry

Abel, E. W.; Stone, F. G. A.; Wilkinson, G., Eds., *Comprehensive Organometallic Chemistry II*, Pergamon: New York, 1995. A review of all of organometallic chemistry. Several chapters describe arene complexes of transition metals.

Collman, J. P.; Hegedus, L. S.; Norton, J. R.; Finke, R. G. *Principles and Applications of Organotransition Metal Chemistry*, University Science Books: Mill Valley, CA, 1987. A survey of organometallic compounds and their use as catalysts.

Elschenbroich, Ch.; Salzer, A. *Organometallics, A Concise Introduction*, 2nd ed., VCH: New York, 1992. An introductory survey of the subject.

Komiya, S., Ed., *Synthesis of Organometallic Compounds*, Wiley: New York, 1997. A survey of the synthesis and applications of selected organometallic compounds.

Organoiron Chemistry: $(C_5H_5)_2Fe_2(CO)_4$ and $(C_5H_5)Fe(CO)_2CH_3$

Note: This experiment requires 8 hours spread over two laboratory periods.

Unlike classical coordination chemistry, organometallic chemistry often involves reactions and compounds that can only be studied in the absence of air and water. You will have an opportunity to learn how to handle air- and water-sensitive compounds in this experiment. The first part of the present experiment examines the preparation of $(C_5H_5)_2Fe_2(CO)_4$ by the reaction of $Fe(CO)_5$ and dicyclopentadiene. The balanced equation is

$$2\,Fe(CO)_5 \;+\; C_{10}H_{12} \;\longrightarrow\; (C_5H_5)_2Fe_2(CO)_4 \;+\; 6\,CO \;+\; H_2$$

In this synthesis, the dicyclopentadiene cracks to give cyclopentadiene (see Experiment 20), which then reacts with $Fe(CO)_5$. The product contains an iron–iron single bond (see Experiment 11 for a discussion of metal–metal bonding). The structure and properties of this dinuclear complex illustrate a number of useful principles as discussed below.

In solution, the compound $(C_5H_5)_2Fe_2(CO)_4$ exists in three isomeric forms; these are distinguished by the relative arrangement of the ligands (Fig. 17-1). The cis and trans isomers are most abundant; they differ in the relative positions of the C_5H_5 ligands. In both of the cis and trans isomers, two of the four CO ligands are terminal whereas the other two CO ligands bridge between the iron atoms. The third and least abundant isomer has no bridging ligands so that the metals are held together only by the Fe–Fe bond.

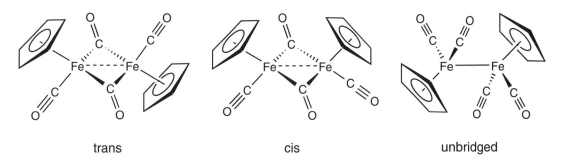

trans   cis   unbridged

Figure 17-1
Structures of the three isomers of $(C_5H_5)_2Fe_2(CO)_4$.

171

In solution, the three isomers of $(C_5H_5)_2Fe_2(CO)_4$ interconvert. Molecules that exhibit such interconverting structures are called *fluxional*. The fluxional processes for $(C_5H_5)_2Fe_2(CO)_4$ are so fast that its proton nuclear magnetic resonance (^1H NMR) spectrum shows only a single time-averaged signal for the C_5H_5 ligands. The fluxional process occurs by movement of the bridging CO ligands to terminal positions to give the unbridged isomer, followed by reformation of the bridging CO ligands.

The fluxional process is not fast enough to average the infrared (IR) absorptions, however, and instead separate absorptions are seen for each isomer. For example, for both the cis and trans isomers of $(C_5H_5)_2Fe_2(CO)_4$, the bridging CO ligands are readily distinguished from the terminal CO ligands by IR spectroscopy. The ν_{CO} bands for the bridging CO ligands are found near 1780 cm^{-1}, that is, about 200 cm^{-1} lower in frequency than the ν_{CO} bands for the terminal CO ligands.

The complex $(C_5H_5)_2Fe_2(CO)_4$ is an important organometallic compound primarily because it serves as a starting material for a wide variety of other organoiron compounds. Halogens cleave the Fe–Fe bond to give halides $(C_5H_5)Fe(CO)_2X$, where X = Cl, Br, or I. The halide ligands in these complexes can be displaced by nucleophiles such as Grignard reagents to give products with Fe–C bonds.

The compound $(C_5H_5)_2Fe_2(CO)_4$ can be reduced with alkali metals and other reagents to give salts $M^+(C_5H_5)Fe(CO)_2^-$. These reduced compounds are powerful nucleophiles. The anion reacts with a wide range of main group and organic electrophiles. In the present experiment, you will generate $K(C_5H_5)Fe(CO)_2$ and alkylate it with methyl iodide to give $(C_5H_5)Fe(CO)_2CH_3$. The reactions are as follows:

$$(C_5H_5)_2Fe_2(CO)_4 \; + \; 2\,KBH(C_2H_5)_3$$

$$\longrightarrow \; 2\,K(C_5H_5)Fe(CO)_2 \; + \; H_2 \; + \; 2\,B(C_2H_5)_3$$

$$K(C_5H_5)Fe(CO)_2 \; + \; CH_3I \; \longrightarrow \; (C_5H_5)Fe(CO)_2CH_3 \; + \; KI$$

Schlenk Line Techniques

In this experiment, you will become acquainted with a number of techniques used to handle air-sensitive compounds. For many compounds the strict observance of these techniques is essential to complete the synthesis successfully. Even if the final compound in a synthesis is air stable, it is sometimes the case that an intermediate is very air sensitive. For example, in the second part of this experiment, the intermediate $K(C_5H_5)Fe(CO)_2$ is extremely reactive toward O_2 and H_2O whereas the product $(C_5H_5)Fe(CO)_2CH_3$ is stable in air for several hours. In the following paragraphs, a few of the most basic techniques for handling air-sensitive compounds are discussed.

The first step in the synthesis of an air-sensitive compound is to establish an inert atmosphere. Reaction vessels can be evacuated with a vacuum pump and then refilled with an inert gas. This vac–refill method is best accomplished with an apparatus called a double manifold (Fig. 17-2). This apparatus consists of

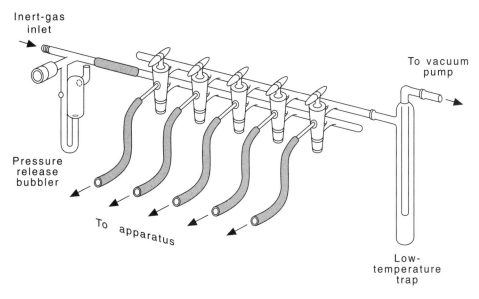

Inert-gas inlet

To vacuum pump

Pressure release bubbler

To apparatus

Low-temperature trap

Figure 17-2
Double manifold used for handling air-sensitive compounds. (Figure adapted from Shriver, D. F.; Drezdzon, M. A. *The Manipulation of Air-Sensitive Compounds*, 2nd ed., Wiley: New York, 1986.)

two glass tubes (called manifolds), one that is connected to a source of inert gas (typically N_2) and the other that is connected to a vacuum pump typically capable of reducing the pressure to below 10^{-2} Torr (see Experiment 6 for the definition of the Torr unit). The N_2 manifold is connected to a mercury bubbler (see below), and the vacuum manifold usually has a trap cooled with liquid N_2. The trap prevents material in the reaction flask from contaminating the pump oil and vice versa. Reaction flasks are connected to the double manifold by means of special stopcocks, which allow one to switch easily from vacuum to the inert gas. The fraction of air remaining in a flask after several vac–refill processes can be described by the expression $(f)^n$, where f is the fraction of the original atmosphere remaining after one vac–refill cycle, and n is the number of cycles.

To prevent air from being accidentally introduced into the reaction vessel, the pressure inside the inert gas manifold (and thus inside the reaction flask) should be kept above atmospheric pressure by venting the inert gas in the manifold through a mercury bubbler. The bubbler design and amount of mercury added to the bubbler determine the amount of positive pressure in the manifold. The use of positive pressure, however, can cause joints or stoppers to come loose unless they are secured with clips, rubber bands, or wires.

If a double manifold is not available, reaction vessels can also be filled with an inert atmosphere simply by flushing them with the inert gas. This method does not require a vacuum pump, but it is not as easy to determine when the atmosphere inside the vessel is ready to use. Flushing is most effective when the inert gas enters at one end of the apparatus and exits at the other. You can use this method in the present experiment because the apparatus is fairly simple.

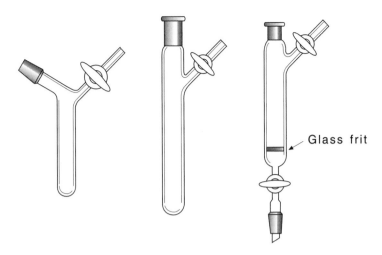

Glass frit

Figure 17-3
Two types of Schlenk flasks and a Schlenk filter. (Courtesy of Kimble/Kontes Glass Inc.)

Air-sensitive compounds are conveniently handled in Schlenk flasks. The basic Schlenk flask is simply a glass vessel fitted with a ground glass joint and a side arm equipped with a stopcock (Fig. 17-3). This design allows one to introduce reagents into and withdraw reagents from the flask while providing a protective atmosphere through the side arm. Schlenk flasks are often used in conjunction with syringes and septa. Rubber septa are available in a variety of sizes, and it is possible to find one that will fit tightly into any joint. Liquid reagents can be syringed into the flask through the rubber septum. Larger volumes of liquid are often added using a cannula, which is a flexible double tipped needle usually made of stainless steel or a chemically inert plastic. Schlenk filter flasks are used to filter air sensitive solutions, in order to remove undesirable solids or to collect a product. An alternative filtration technique involves the use of a "filter cannula," in which one end of a cannula is fitted with a filter on its tip. Consult Shriver and Drezdzon's book for more information about septum and cannula techniques.

Infrared Spectra of Solutions
As discussed in Experiment 19, IR spectra can be measured for a variety of sample types, including solids, gases, and liquids. It is most common to record spectra of solids, but there are situations in which it is advantageous to study solutions. The IR absorption peaks seen for solutions are usually narrower than those seen for solids. Furthermore, solution spectra are convenient for monitoring the progress of a reaction because small samples can easily be removed from an ongoing reaction with a syringe. Solution spectra can also be used to determine whether the structure of a complex changes upon dissolution. For example, crystals of the compound $(C_5H_5)_2Fe_2(CO)_4$ contain only one of the three isomeric forms shown in Figure 17-1, but in solution all three isomers coexist. It is not uncommon for compounds to adopt different structures in solution than in the solid state.

It is relatively easy to measure a solution IR spectrum. Apart from the obvious fact that the sample must be soluble in some solvent, the main requirement is that the IR bands of the solvent should not overlap with those due to the solute. Metal carbonyl compounds are ideally suited for analysis by solution IR spectroscopy because the C–O stretching bands occur in a region of the spectrum where few organic solvents absorb, between 1750 and 2100 cm^{-1}. Furthermore, these ν_{CO} bands are extremely intense, thus allowing one to record spectra even from very dilute samples.

The resolution of IR spectra is strongly affected by the choice of solvent. For example, the C–O stretching bands of metal carbonyls are typically broader when polar solvents are used. This broadening results from the strong interaction between polar solvents and the vibrating CO ligand. Because the chemist's ability to distinguish between samples is diminished if the absorption bands are too broad, it is desirable to use nonpolar solvents such as cyclohexane and carbon disulfide. Fortunately, many metal carbonyls are soluble in nonpolar solvents. Even if it is not possible to use such nonpolar solvents, the solution IR technique is still very useful.

Solution IR spectra are measured from solutions that are placed in a cell of the type shown in Figure 17-4. For organic solvents, it is common to employ a cell made from NaCl. If one needs to record the IR spectrum of an aqueous solution, it is common to use a cell made of AgCl. (Transmission ranges of various materials are given in Appendix 3.) Be certain that the solvent is compatible with the cell being used. Water and certain alcohols dissolve NaCl whereas amines and sulfur compounds are incompatible with AgCl cells.

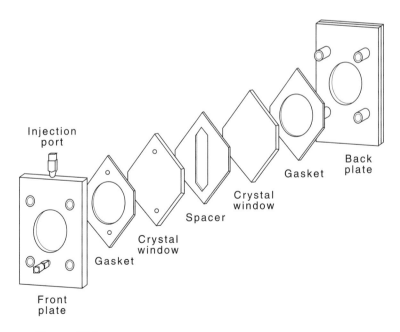

Figure 17-4
Solution IR cell. (Courtesy of International Crystal Laboratories.)

After choosing a solvent and a suitable IR cell, the first step is to record the IR spectrum of the pure solvent. If there are no strong bands in the region of interest, then the solvent is suitable. Once the solvent spectrum has been recorded, record the spectrum of your compound dissolved in the same solvent. If desired, peaks due to the solvent can be removed from the latter spectrum by subtracting the spectrum of the pure solvent.

EXPERIMENTAL PROCEDURE

Cyclopentadienyl(dicarbonyl)iron dimer, $(C_5H_5)_2Fe_2(CO)_4$

Safety note: Iron pentacarbonyl is a toxic liquid with a high vapor pressure and a musty odor. The relatively high volatility of $Fe(CO)_5$ makes it particularly dangerous, and it should be stored and handled in a hood. All of its reactions involve the evolution of CO and this is a further reason for conducting all syntheses in an efficient hood. Reactions of $Fe(CO)_5$ often afford as a side product finely divided iron powder; this powder can inflame in air. The compounds prepared in this experiment can be filtered in air but you should avoid drawing air through the residues after filtration. They should be dried in a vacuum instead.

In a hood, assemble the reaction vessel as shown in Figure 17-5. Flush the

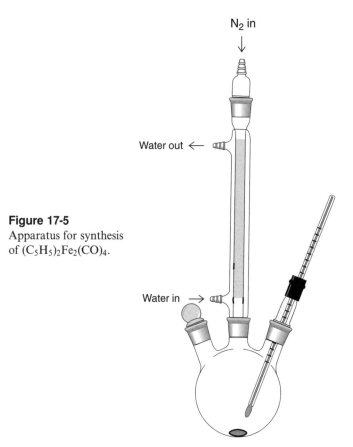

N₂ in

Water out ←

Water in →

Figure 17-5
Apparatus for synthesis
of $(C_5H_5)_2Fe_2(CO)_4$.

vessel for 5 min with a gentle stream of N_2 by removing the thermometer and adjusting the flow of gas. Then add 64 mL (60 g, 455 mmol) of dicyclopentadiene ($C_{10}H_{12}$) to the round-bottom flask. To minimize exposure to $Fe(CO)_5$, handle it in a hood and use a syringe to measure out and add 10 mL (14.6 g, 70.5 mmol) of $Fe(CO)_5$ into the flask. The constant stream of N_2 will prevent air from entering the flask while the reactants are being added. After replacing the thermometer, turn off the N_2 flow both at the tank and at the inlet on the flask. Make sure, however, that the gas inside the flask can vent out through the bubbler.

Heat the reaction mixture at 135 °C for 8–16 h. Do not let the temperature go below 130 or above 140 °C. Below 130 °C very little reaction occurs, and above 140 °C pyrophoric, finely divided metallic iron is produced. *Here is a good stopping point for the first day's work*; leave the solution refluxing and arrange to meet your instructor tomorrow to turn the heat off.

When the reaction solution has cooled, collect the red-violet crystals of $(C_5H_5)_2Fe_2(CO)_4$ on a suction filter (Fig. 13-1), and wash them four times with 20-mL portions of hexane (or a low-boiling petroleum fraction). Dry the product in a vacuum. Record the yield. Measure the IR spectrum of the product as a $CHCl_3$ solution. Its ^1H NMR spectrum (Experiment 16) exhibits a single sharp resonance at δ 4.64.

Generally, the crude product can be used successfully to prepare the $(C_5H_5)Fe(CO)_2CH_3$ compound below. If you wish to purify the product anyway, dissolve it in a minimum volume of CH_2Cl_2 and filter this solution, rinsing the frit with a little CH_2Cl_2. Transfer the filtrate into a round-bottom flask and then dilute it with an equal volume of hexane. Slowly evaporate the solvent under a water aspirator vacuum, ideally using a rotary evaporator, to about one-half of the original volume. The more volatile CH_2Cl_2 will evaporate readily, causing the product to crystallize.

Solution Infrared Spectrum

Dissolve a few crystals of your sample in about 2 milliliters of chloroform and use a syringe or pipet to transfer some of this solution to the IR sample cell (Fig. 17-4). Add enough solution so that it fills the gap between the two windows. If necessary, cap the filling ports so that the solvent does not leak out. Place the sample cell in the IR spectrometer and record the spectrum between 1600 and 2500 cm^{-1}. If you are using a grating IR instrument, place a matching cell containing only solvent in the reference beam of the spectrophotometer. If you are using a Fourier transform instrument, record the spectrum of the matching cell containing only solvent and use this spectrum to correct the background. If the sample and reference cells are not well matched, solvent absorptions may appear in the spectrum. After recording the spectrum, pour the solution out of the cell. Clean the cell by filling it with solvent and decanting. The washing procedure should be repeated at least one more time. After drying the cells with a rapid flush of dry air or nitrogen, return them to the desiccator. If sample solution is spilled on the outside surface of the cell window, rinse it off with solvent. Otherwise, subsequent samples measured in that cell will always give a spectrum of your compound.

Cyclopentadienyl(dicarbonyl)methyliron, $(C_5H_5)Fe(CO)_2CH_3$

Safety note: methyl iodide is a suspected carcinogen. Solutions of KBHEt$_3$ are toxic and inflammable.

Clamp a 100-mL three-neck round-bottom flask on a ring stand and attach a reflux condenser topped with a N_2 gas inlet to the central arm. Flush the flask with a gentle flow of N_2 for one minute. Place a greased stopper in one of the side necks and fit the other with a rubber septum. Add a magnetic stir bar followed by 1 g (2.8 mmol) of $(C_5H_5)_2Fe_2(CO)_4$ and 20 mL of dry tetrahydrofuran (THF) through the side neck of the flask. It helps to turn up the N_2 flow to prevent air from entering the flask during this operation, but do not turn the gas flow so high that it blows the iron complex out of the neck! Replace the septum and heat the solution to reflux with a rheostat-controlled heating mantle. After the solution comes to reflux, use a syringe to inject (through the septum) 6 mL of a 1 M solution of $KBH(C_2H_5)_3$ in THF (6 mmol). Turn off the heat, and let the solution cool slowly to room temperature over 30 min. Note any color changes that occur as the reduction proceeds. Clean the syringe by rinsing it with some hexane.

To the THF solution of $K(C_5H_5)Fe(CO)_2$, add 0.4 mL (6.4 mmol) of methyl iodide with a syringe. After a few minutes, place a gas inlet in one of the side arms, and connect the gas inlet to a source of N_2. While flushing the flask with N_2, replace the condenser with a glass stopper. Evaporate the solution to dryness by attaching the flask to a vacuum line or water aspirator.

The next step involves purification of $(C_5H_5)Fe(CO)_2CH_3$ by sublimation. Connect the flask containing the crude product to a Schlenk line and evacuate the tubing while keeping the stopcock on the flask closed. Then refill the tubing with N_2 and open the stopcock on the flask. Remove the central glass stopper and replace it with a water-cooled sublimer probe (see Fig. 20-3). Connect the sublimer to hosing and flow cold water through it. Because the product is so volatile, it can easily sublime into the vacuum pump. To prevent this, open the flask to the vacuum only long enough to establish a vacuum. Usually, this takes about a minute. Then close the stopcock on the flask, and warm the flask to about 50 °C with a hot water bath. If very little sublimate appears on the sublimation probe, briefly reevacuate the flask. The orange waxy product may be removed from the sublimation probe by scraping it into a small Schlenk flask for storage. This operation can be conducted in air if it is done quickly.

Calculate the yield of your product. Record its IR spectrum as a solution in chloroform, and record its ^1H NMR spectrum in CHCl$_3$ or CDCl$_3$ (ask your instructor which of these two solvents you should use). See Experiment 7 for information about preparing NMR samples.

REPORT

Include the following:
1. Percentage yield for both compounds.
2. IR spectrum of a solution of $(C_5H_5)_2Fe_2(CO)_4$ in CHCl$_3$. Assign the C–O stretching bands you see.
3. ^1H NMR spectrum of $(C_5H_5)Fe(CO)_2CH_3$. Include assignments of the peaks.

PROBLEMS

1. The preparations in this experiment were conducted in a nitrogen atmosphere. What products would probably have formed if the reactions had been carried out in air?

2. Discuss the bonding in $(C_5H_5)_2Fe_2(CO)_4$ and account for the diamagnetism of this complex.

3. Predict the structure of the product formed from the reaction of $SnCl_2(C_4H_9)_2$ with excess $K(C_5H_5)Fe(CO)_2$.

4. Although both cis and trans isomers of $(C_5H_5)_2Fe_2(CO)_4$ are known to be present in solution, the ^1H NMR spectrum at $25\,°C$ shows only a sharp singlet. Why?

INDEPENDENT STUDIES

A. Prepare $(C_5H_5)_2Mo_2(CO)_6$ and study its reactions. (Lucas, C. R.; Walsh, K. A. *J. Chem. Educ.* **1987**, *64*, 265.)

B. Obtain the mass spectrum of $(C_5H_5)_2Fe_2(CO)_4$ and make assignments of all peaks.

C. Prepare and characterize $(C_5H_5)Fe(CO)_2Cl$ or $(C_5H_5)Fe(CO)_2I$ obtained from $(C_5H_5)_2Fe_2(CO)_4$. (Fischer, E. O.; Moser, E. *Inorg. Synth.* **1970**, *12*, 35. King, R. B.; Stone, F. G. A. *Inorg. Synth.* **1963**, *7*, 110.)

D. Prepare the tetranuclear cluster $(C_5H_5)_4Fe_4(CO)_4$ from $(C_5H_5)_2Fe_2(CO)_4$. (Westmeyer, M. D.; Massa, M. A.; Rauchfuss, T. B.; Wilson, S. R. *J. Am. Chem. Soc.* **1998**, *120*, 114.)

E. Prepare $(C_5H_5)RuCl[P(C_6H_5)_3]_2$ and study its reactions. (Ballester, L.; Gutiérrez, A.; Perpiñán, M. F. *J. Chem. Educ.* **1989**, *66*, 777.)

REFERENCES

$(C_5H_5)_2Fe_2(CO)_4$ and $(C_5H_5)Fe(CO)_2CH_3$

Bryan, R. F.; Greene, P. T. *J. Chem. Soc. (A)* **1970**, 3064. Kirchner, R. M.; Marks, T. J.; Kristoff, J. S.; Ibers, J. A. *J. Am. Chem. Soc.* **1973**, *95*, 6602. X-ray structural studies of *trans*-$(C_5H_5)_2Fe_2(CO)_4$.

Bryan, R. F.; Greene, P. T.; Newlands, M. J.; Field, D. S. *J. Chem. Soc. (A)* **1970**, 3068. X-ray structural study of *cis*-$(C_5H_5)_2Fe_2(CO)_4$.

Bullett, J. G.; Cotton, F. A.; Marks, T. J. *Inorg. Chem.* **1972**, *11*, 671. Solution structure of $(C_5H_5)_2Fe_2(CO)_4$.

Gladysz, J. A.; Williams, G. M.; Tam, W.; Johnson, D. L.; Parker, D. W.; Selover, J. C. *Inorg. Chem.* **1979**, *18*, 553. Reduction of metal carbonyls with trialkylborohydride reagents.

King, R. B. *Organometallic Syntheses*, Academic: New York, 1965; Vol. 1. Older but still classic collection of syntheses of organotransition metal compounds.

Mahmoud, K. A.; Rest, A. J.; Alt, H. G. *J. Chem. Soc., Dalton Trans.* **1985**, 1365. Photochemistry of $(C_5H_5)Fe(CO)_2CH_3$.

Osella, D.; Gambino, O.; Nervi, C.; Stein, E.; Jaouen, G.; Vessières, A. *Organometallics* **1994**, *13*, 3110. Estrogen derivatives of transition metal carbonyls for enhanced analytical detection.

Pearson, A. J. *Iron Compounds in Organic Synthesis*, Academic: New York, 1994.

Piper, T. S.; Wilkinson, G. *J. Inorg. Nucl. Chem.* **1956**, *3*, 104. Original preparations of $(C_5H_5)_2Fe_2(CO)_4$ and $(C_5H_5)Fe(CO)_2CH_3$.

Schlenk Line Methods

Errington, R. J., Ed. *Advanced Practical Inorganic and Metalorganic Chemistry*, Blackie: New York, 1997.

Shriver, D. F.; Drezdzon, M. A. *The Manipulation of Air-sensitive Compounds*, 2nd ed., Wiley: New York, 1986.

Vosejpka, L. J. S. *J. Chem. Educ.* **1993**, *70*, 665. Transfer of air-sensitive liquids without a double manifold.

For References to Infrared Spectroscopy, see Experiment 19.

The Metal Carbonyl Cluster $Fe_3(CO)_{12}$

Note: This experiment requires 6 hours spread over two laboratory periods.

Before the 1960s, compounds in which two or more transition elements formed bonds with one another were rare. Since then, however, the number of compounds with metal–metal bonds has increased dramatically. The simplest are *dinuclear* complexes in which two metals are linked solely by metal–metal bonds. Examples include $[Mn(CO)_5]_2$ and $[(C_5H_5)_2Os]_2^{2+}$. The presence of the metal–metal bond is strongly indicated by two facts: the compounds are diamagnetic and they do not easily dissociate into two pieces. In many other complexes, two transition elements are held together both by a metal–metal bond and by one or more ligands that bridge between the two metals (see Experiments 11 and 17). In such cases, it is sometimes difficult to establish whether the two metals are held close together by the bridging ligands, a metal–metal bond, or both.

Compounds that contain three or more metals held together by metal–metal bonds are called *clusters*. The number of metal atoms in a cluster varies widely; one of the largest clusters known has the formula $Cu_{146}Se_{73}[P(C_6H_5)_3]_{30}$. Most clusters, however, have six metals or fewer. All transition metals form cluster compounds; the structures of two examples are shown below:

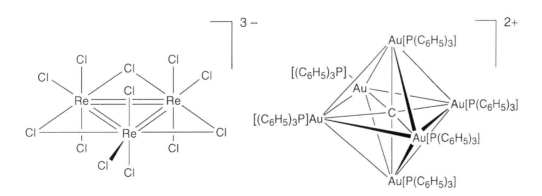

181

Metal clusters undergo all of the reactions characteristic of mononuclear complexes. In addition, they sometimes display distinctive reactions arising from the simultaneous coordination of ligands at adjacent metal centers.

Most metal clusters violate the simple electron counting rules developed for mononuclear species. A fairly successful approach to understanding structural patterns in many metal–metal bonded compounds, especially metal clusters, builds on our knowledge of boron hydrides (Experiment 4). For example, one surprising but useful observation is that BH and $Fe(CO)_3$ are often interchangeable vertices in clusters. In general, theoretical studies of metal clusters lead to two conclusions.

1. The highest energy electrons in metal clusters are those associated with the metal–metal bonds.
2. The interactions between metals often cannot be described in terms of two-center, two-electron bonds.

The first statement tells us that oxidizing or reducing a metal cluster will most strongly affect the M–M interactions. The second statement reminds us that, as in boron hydrides, the lines drawn between atoms in clusters do not necessarily imply the presence of electron pairs.

In this experiment, you will prepare one of the first complexes known to contain a metal–metal bond, $Fe_3(CO)_{12}$. The synthesis proceeds in two steps:

$$3 \ Fe(CO)_5 + H_2O + N(C_2H_5)_3$$

$$\longrightarrow [N(C_2H_5)_3H][HFe_3(CO)_{11}] + CO_2 + 3 \ CO$$

$$12 \ [N(C_2H_5)_3H][HFe_3(CO)_{11}] + 21 \ HCl$$

$$\longrightarrow 11 \ Fe_3(CO)_{12} + 10.5 \ H_2 + 12 \ [N(C_2H_5)_3H]Cl + 3 \ FeCl_3$$

The first step begins with the addition of hydroxide to a carbonyl ligand to give the unstable formate complex $Fe(CO)_4(CO_2H)^-$. The formate group loses CO_2 to afford the hydride complex $HFe(CO)_4^-$ which, in a series of steps, combines with two equivalents of $Fe(CO)_5$ to give the hydrido cluster $HFe_3(CO)_{11}^-$ and CO. The presence of a bridging hydride ligand in $HFe_3(CO)_{11}^-$ should remind you of the similarity between the chemistry of boron hydrides and metal clusters. The conversion of $HFe_3(CO)_{11}^-$ to $Fe_3(CO)_{12}$ is completed by the addition of hydrochloric acid, which produces H_2, $FeCl_3$, and the desired product. The structures of $HFe_3(CO)_{11}^-$ and $Fe_3(CO)_{12}$ are shown below:

The CO ligands in $Fe_3(CO)_{12}$ occupy both terminal and bridging positions (see the discussion in Experiment 17). In contrast, the CO ligands in the related molecule $Ru_3(CO)_{12}$ all occupy terminal positions.

Mass Spectrometry

Mass spectrometry is an important method for the characterization of inorganic and organometallic compounds. One advantage of the technique is that the amount of sample required is very small (micrograms). This technique is especially useful for characterizing metal carbonyl clusters because it is often possible to count the number of carbonyl ligands and metal atoms by analyzing how the cluster undergoes fragmentation.

A schematic diagram of a mass spectrometer is shown in Figure 18-1:

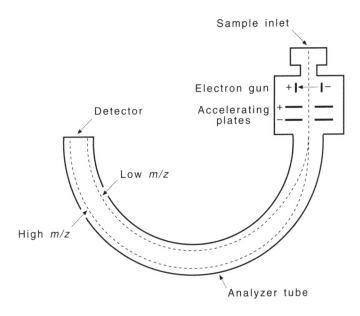

Figure 18-1
Schematic drawing of a mass spectrometer.

The entire system is under high vacuum (10^{-8} Torr); see Experiment 6 for a definition of the Torr unit. For many instruments the sample must either be a gas or be sufficiently volatile to be converted to a gas at elevated temperatures. Metal carbonyls are especially well suited for this technique because they are usually somewhat volatile and thermally robust. The high sensitivity of the instrument allows samples with exceedingly low volatilities to be examined. Ionic compounds and polymers are two of the few types of materials that are more challenging to study by mass spectrometry, but the recent development of "soft" ionization techniques has changed this situation, and the mass spectrum of virtually any compound can now be measured.

In electron impact mass spectrometry, the sample is introduced into the inlet system and diffuses into the ionization chamber, where it is bombarded by a beam of electrons from an electron gun. If the electrons are sufficiently energetic, they may simply ionize molecules of the sample. The resulting ions will diffuse to the accelerating plates, where they will be given a known kinetic energy by the known voltage between these plates. The accelerated ions then pass between the poles of a magnet. Because a magnetic field bends the path of a moving charged particle, the ions follow curved paths, as shown in Figure 18-1.

The radius of curvature of the ion path is determined by the ion's mass (m) and charge (z). Light-weight ions and those with high charge will follow highly curved paths because they are most strongly affected by the magnetic field. In other words, the radius of curvature will be smallest for ions with a low m/z ratio and greatest for those with a high m/z ratio. For a group of $+1$ ions, the radius of curvature and the position at which they strike the detector will only depend on the mass of the ion. The accuracy of mass measurements by this technique is determined by the nature of the instrument, but in most cases ions whose masses differ by 0.5 atomic mass unit (amu) are easily distinguished.

Consider the mass spectrum of $Ni(CO)_4$ (Fig. 18-2). The initial ionization process produces $Ni(CO)_4{}^+$ ions. These $Ni(CO)_4{}^+$ ions are called *parent ions*, and ions of lower mass generated by fragmentation of the parent ion are called *daughter ions*. The total mass of the $Ni(CO)_4{}^+$ ion containing the ^{58}Ni isotope is 170, assuming the remainder of the molecule to contain only ^{12}C and ^{16}O isotopes. Nickel, however, exists as several stable isotopes with natural abundances as follows:

Isotope	Natural Abundance (%)
^{58}Ni	67.77
^{60}Ni	26.16
^{61}Ni	1.25
^{62}Ni	3.66
^{64}Ni	1.16

Therefore, $Ni(CO)_4{}^+$ ions that contain other Ni isotopes will also be present, and peaks with $m/z = 170, 172, 173, 174$, and 176 will appear in the mass spectrum. Their relative peak heights will be proportional to the relative abundances of the Ni isotopes. Thus, in the spectrum shown in Figure 18-2, the most intense peak of the parent ion $Ni(CO)_4{}^+$ occurs at a mass of 170 and the next most intense peak is that of mass 172.

To this point, it has been assumed that the only carbon and oxygen isotopes present are ^{12}C and ^{16}O. For O, this is essentially true because 99.76% of naturally occurring oxygen is ^{16}O. For C, this assumption is less valid:

Isotope	Natural Abundance (%)
^{12}C	98.89
^{13}C	1.11

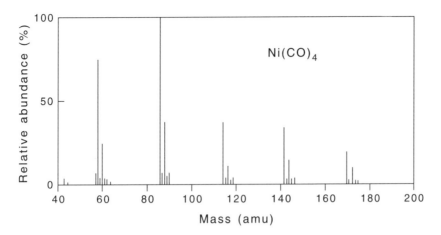

Figure 18-2
Mass spectrum of Ni(CO)₄.

The measurable abundance of ^{13}C means that measurable amounts of $Ni(^{12}CO)_3(^{13}CO)^+$ ($m/z = 171$) should be visible in the mass spectrum. Because 1.1% of the C is ^{13}C, there is a 4.4% probability that one of the four C atoms in $^{58}Ni(CO)_4$ is ^{13}C. In other words, 4.4% of $^{58}Ni(CO)_4$ is $^{58}Ni(^{12}CO)_3(^{13}CO)$. The probability of finding two ^{13}CO groups in the same $^{58}Ni(CO)_4$ molecule is very small, about 0.07%, and consequently the peak due to $^{58}Ni(^{12}CO)_2(^{13}CO)_2$ is usually not seen.

As is obvious from Figure 18-2, a mass spectrum contains more than just parent ion peaks. If the bombarding electrons in the spectrometer are of sufficiently high energy, fragmentation of the molecule also occurs. For example,

$$Ni(CO)_4 + e^- \longrightarrow Ni(CO)_4^+ + 2\,e^-$$

$$Ni(CO)_4 + e^- \longrightarrow Ni(CO)_3^+ + CO + 2\,e^-$$

$$Ni(CO)_4 + e^- \longrightarrow Ni(CO)_2^+ + 2\,CO + 2\,e^-$$

$$Ni(CO)_4 + e^- \longrightarrow Ni(CO)^+ + 3\,CO + 2\,e^-$$

$$Ni(CO)_4 + e^- \longrightarrow Ni^+ + 4\,CO + 2\,e^-$$

and also

$$Ni(CO)_4 + e^- \longrightarrow Ni(CO)_3 + CO^+ + 2\,e^-$$

All of the ions containing Ni, with their characteristic isotopic mass distributions, are found in the spectrum.

In addition to the +1 ions, +2 ions are sometimes observed, but they are generally of low abundance. For example, the +2 ion, $^{58}Ni(^{12}CO)_4^{2+}$ should occur at an m/z value of $170/2 = 85$. The spectrum shows no peak at that mass, which suggests that $Ni(CO)_4^{2+}$ is less easily formed than the other observed ions.

For $^{58}Ni(^{12}CO)_2{}^{2+}$, a peak at $114/2 = 57$ does in fact appear. Likewise, the presence of $^{58}Ni(^{12}CO)^{2+}$ is indicated by a peak at $86/2 = 43$. Like the $+1$ ions, the $+2$ ions also give characteristic isotopic patterns.

By such an analysis as we have made here for $Ni(CO)_4$, it is possible to account for all the peaks in the mass spectrum of $Fe_3(CO)_{12}$. The distribution of Fe isotopes is as follows:

Isotope	Natural Abundance (%)
^{54}Fe	5.82
^{56}Fe	91.66
^{57}Fe	2.19
^{58}Fe	0.33

Because there are many more C atoms in $Fe_3(CO)_{12}$, the abundance of ions containing one ^{13}C atom will also be higher than observed for $Ni(CO)_4$. The natural abundances of the isotopes of Fe and other elements are listed in Appendix 7.

EXPERIMENTAL PROCEDURE

Safety note: Iron pentacarbonyl is a toxic liquid with a high vapor pressure and a musty odor. The relatively high volatility of $Fe(CO)_5$ makes it particularly dangerous, and it should be stored and handled in a hood. All of its reactions involve the evolution of CO and this is a further reason for conducting all syntheses in an efficient hood. Reactions of $Fe(CO)_5$ often afford as a side product finely divided iron powder; this powder can inflame in air. *Caution!* The compounds prepared in this experiment may burn sponaneously in air if finely divided. Treat them with care.

Triethylammonium Hydrido(undecacarbonyl)triiron, $[N(C_2H_5)_3H][HFe_3(CO)_{11}]$

Charge a 100-mL two-neck round-bottom flask with 24 mL of N_2-purged H_2O and a magnetic stir bar. Fit the flask with a reflux condenser topped with a gas inlet and place the flask in an oil bath mounted on a stirring hot plate (a heating mantle may be used in place of the oil bath but stirring will be more difficult). Use silicone grease to lubricate the joints. The gas inlet should be connected with tubing to a "T piece"; the other two arms of the T should be connected to a source of N_2 gas and a bubbler. Flush the flask with nitrogen for a minute, and then stopper the second neck on the flask. Heat the water to boiling, and then immediately allow the water to cool to room temperature, being careful to maintain the reaction flask under an atmosphere of N_2. Under a gentle positive pressure of N_2, remove the stopper in the flask, degrease the joint, and add 11 mL (16 g, 0.082 mol) of $Fe(CO)_5$ followed by 8.3 mL (6 g, 0.059 mol) of triethylamine. Replace the stopper, and with the flask open to the N_2 source and bubbler, heat the solution to 80 °C overnight. *Note*: If the reaction temperature drops below 75 °C the reaction will be too slow, and if it rises above 90 °C the products will decompose. Keep the temperature between these two points. It is helpful to use a temperature controller and a thermocouple, if these are available, to regulate the temperature overnight. *Here is a good stopping point for the first day's work.* Make arrangements with your instructor to turn off the heat tomorrow.

Dodecacarbonyltriiron, Fe$_3$(CO)$_{12}$

A red-brown oil of [N(C$_2$H$_5$)$_3$H][HFe$_3$(CO)$_{11}$] should appear after the reaction mixture has been heated at 80 °C overnight. Cool the reaction mixture to room temperature. The red-brown oil of [N(C$_2$H$_5$)$_3$H][HFe$_3$(CO)$_{11}$] solidifies upon cooling. Filter off this crude salt in air and wash it with 200 mL of water. *Note:* Some care should be exercised here because unchanged Fe(CO)$_5$ may still remain. Treat the washings with some bromine water to destroy any carbonyl compounds.

Record the IR spectrum of the collected solid product as a KBr pellet. If it is necessary to stop at this stage, the sample should be stored under an inert atmosphere.

To convert the salt to Fe$_3$(CO)$_{12}$, add the solid [N(C$_2$H$_5$)$_3$H][HFe$_3$(CO)$_{11}$] to a 200-mL round-bottom flask containing a stir bar. Next add 60 mL of methanol to dissolve the salt. *Cautiously* add a solution of 50 mL of H$_2$O and 50 mL of concentrated HCl. Heat this solution for about 30 min at reflux. The solution should become slightly greenish and a black crystalline solid should appear in the flask (much if it will float on top of the solution). Collect the product by filtration and wash it with 10 mL each of water, methanol, and hexane. Minimize the amount of air drawn through the solid, and instead dry the product in vacuum. Record the yield and measure the melting point, the IR spectrum, and (if possible) the mass spectrum of the product. If a mass spectrometer is not available for use with your sample, analyze the mass spectrum of Fe$_3$(CO)$_{12}$ given in Appendix 10.

REPORT

Include the following:
1. Percentage yield and melting point of Fe$_3$(CO)$_{12}$.
2. IR spectrum and its interpretation in terms of the structure of the product.
3. Mass spectrum of Fe$_3$(CO)$_{12}$ with assignments of the major fragments.

PROBLEMS

1. Write out a mechanism for the reaction of OH$^-$ with Fe(CO)$_5$.
2. The reduction of Fe$_3$(CO)$_{12}$ with Na produces the salt Na$_2$Fe(CO)$_4$. Predict the structure of the anion.
3. The photolysis of Fe(CO)$_5$ affords Fe$_2$(CO)$_9$. Predict the structure of the latter species.
4. Calculate the isotope pattern for the molecular ion of Fe$_3$(CO)$_{12}$. See the web site at http://www.shef.ac.uk/chemistry/chemputer/.
5. You observe peaks in the mass spectrum of Fe$_3$(CO)$_{12}$ at $m/z = 448$. Could this peak be due to the fragment Fe$_2$(CO)$_{12}$? Why or why not? If it is not due to this fragment, suggest another possibility.

INDEPENDENT STUDIES

A. Purify $[N(C_2H_5)_3H][HFe_3(CO)_{11}]$ and characterize it by ^1H NMR and IR spectroscopy.

B. Prepare $Fe_3Te_2(CO)_9$. (Lesch, D. A.; Rauchfuss, T. B. *Inorg. Chem.* **1981**, *20*, 3583.)

C. Prepare $Fe_2(CO)_9$ by the photochemical reaction of $Fe(CO)_5$. (Jolly, W. L. *The Synthesis and Characterization of Inorganic Compounds*, Prentice-Hall: Englewood Cliffs, NJ, 1970, pp 472–474.)

D. Prepare $Ru_3(CO)_{12}$ by the reductive carbonylation of hydrated ruthenium trichloride. (Bruce, M. I.; Jensen, C. M.; Jones, N. L. *Inorg. Synth.* **1989**, *26*, 259.)

E. Purify $Fe_3(CO)_{12}$ by Soxhlet extraction into pentane under a nitrogen atmosphere. (see Experiment 5 for details on Soxhlet extraction.)

F. Prepare $Co_3(CCl)(CO)_9$ and study its chemistry. (Humphrey, M. G.; Rowbottom, C. A. *J. Chem. Educ.* **1994**, *71*, 985.)

REFERENCES

Metal Cluster Compounds

Abel, E. W.; Stone, F. G. A.; Wilkinson, G., Eds. *Comprehensive Organometallic Chemistry II*, Pergamon: New York, 1995. Much of Volume 7 and Chapter 4 of Volume 8 discuss the chemistry of transition metal cluster compounds.

Cotton, F. A.; Wilkinson, G. *Advanced Inorganic Chemistry*, 5th ed., Wiley: NY, 1988. Concise reviews of cluster chemistry with an emphasis on metal halide and metal carbonyl clusters.

Housecroft, C. E. *Metal–Metal Bonded Carbonyl Dimers and Clusters*, Oxford: New York, 1996.

Krautscheid, H.; Fenske, D.; Braun, G.; Semmelmann, M. *Angew. Chem., Int. Ed. Engl.* **1993**, *32*, 1364. Report of the large metal cluster $Cu_{146}Se_{73}[P(C_6H_5)_3]_{30}$.

McFarlane, W.; Wilkinson, G. *Inorg. Synth.* **1966**, *8*, 181. Synthesis of $Fe_3(CO)_{12}$ from $Fe(CO)_5$ via $[N(C_2H_5)_3H][HFe_3(CO)_{11}]$.

Schmidbaur, H. *Pure Appl. Chem.* **1993**, *65*, 691. Polygold clusters containing interstitial atoms.

Shriver, D. F.; Whitmire, K. H. in *Comprehensive Organometallic Chemistry*, Wilkinson, G.; Stone, F. G. A.; Abel, E. W., Eds., Pergamon: New York, 1982; Vol. 4. The chemistry of polynuclear organoiron compounds including carbonyl complexes.

Süss-Fink, G.; Meister, G. *Adv. Organomet. Chem.* **1993**, *35*, 41. Review of organic transformations catalyzed by organometallic clusters.

Vahrenkamp, H. *Adv. Organomet. Chem.* **1983**, *22*, 164. A summary of some elementary reactions for organotransition metal clusters.

Characterization of Metal Carbonyls

Adams, F.; Gijbels, R.; Van Grieken, R., Eds. *Inorganic Mass Spectrometry*, Wiley: New York, 1988.

Braterman, P. S. *Metal Carbonyl Spectra*, Academic: New York, 1975.

Ebsworth, E. A. V.; Rankin, D. W. H.; Cradock, S. *Structural Methods in Inorganic Chemistry*, Oxford University: Oxford, UK 1991. A useful texbook that discusses the application of mass spectrometry to inorganic chemistry.

Geiger, W. E.; Connelly, N. G. *Adv. Organomet. Chem.* **1985**, *24*, 87. Review of the electrochemical properties of organometallic clusters.

Hop, C. E. C. A.; Bakhtiar, R. *J. Chem. Educ.* **1996**, *73*, A162. Application of electrospray mass spectrometry to inorganic compounds.

Microscale Synthesis of Vaska's Complex
IrCl(CO)[P(C$_6$H$_5$)$_3$]$_2$

Note: This experiment requires about 6 hours spread over two laboratory periods.

This experiment will introduce you to the use of microscale techniques, the handling of compounds on a scale of 100 mg or less. One of the advantages of microscale synthesis is the low cost of working with small amounts of reagents and solvents. This consideration is particularly important in this experiment, which uses the expensive metal iridium. Other advantages of microscale techniques are that evaporations and filtrations proceed more quickly than with conventional scale synthesis, and reaction wastes are minimized. You will find that, with practice, microscale syntheses are easier than gram-scale chemistry so long as you are careful. The main problems are mechanical losses due to small amounts of the sample adhering to the apparatus. Thus it is important to minimize the number of transfers and to use small reaction vessels.

Many important concepts in organometallic chemistry relevant to reactivity and catalysis are illustrated by the behavior of the complex prepared in this experiment, *trans*-chlorocarbonylbis(triphenylphosphine)iridium(I), or *trans*-IrCl(CO)[P(C$_6$H$_5$)$_3$]$_2$. This compound is called Vaska's complex in recognition of the researcher who first demonstrated its versatile chemical properties.

Vaska's complex is a square planar complex of iridium(I) with *trans*-triphenylphosphine ligands:

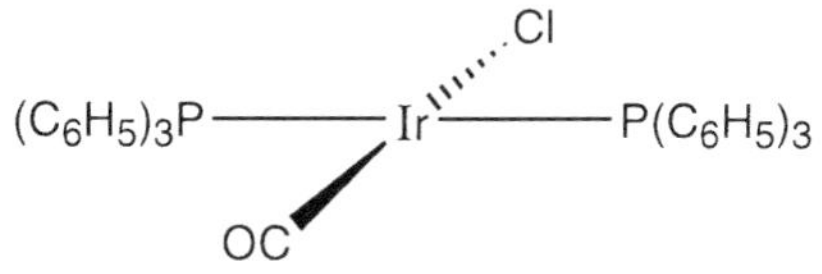

Vaska's complex nicely illustrates the concept that the reactivity of a metal complex is related to its valence electron count. Complexes with 18 valence electrons are called saturated and generally do not bind other ligands. Complexes with 16 valence electrons, such as Vaska's complex, are unsaturated and able to add certain two-electron donors, L, to form adducts:

$$IrCl(CO)[P(C_6H_5)_3]_2 \; + \; L \; \rightleftharpoons \; IrCl(CO)[P(C_6H_5)_3]_2L$$

16 e$^-$ complex   18 e$^-$ complex   189

Vaska's complex also undergoes reactions called oxidative additions in which the Ir^I center inserts into the σ bond of certain molecules. The resulting fragments both become bound to the metal and cause an increase in its formal oxidation state. Vaska's complex undergoes oxidative addition reactions with a large number of simple molecules including conventional oxidants such as Cl_2, strong acids such as HCl, and many other agents such as H_2, CH_3I, $C_6H_5C(O)Cl$, and $(CH_3)_3SiH$. The structures of some of these oxidative addition products are as follows:

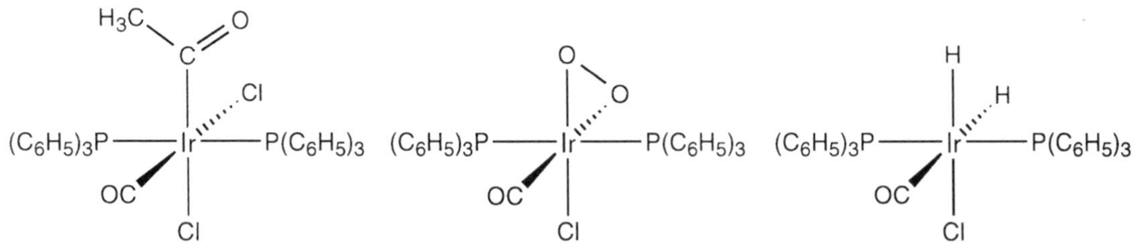

Vaska's complex can be prepared from both of the two commonly available sources of iridium, $IrCl_3 \cdot 3\,H_2O$ and H_2IrCl_6. A possible balanced equation for one of these reactions is

$$H_2IrCl_6 + 3.5\,P(C_6H_5)_3 + HCON(CH_3)_2 + 4\,C_6H_5NH_2 + 1.5\,H_2O \longrightarrow$$

$$IrCl(CO)[P(C_6H_5)_3]_2 + 4\,C_6H_5NH_3{}^+Cl^-$$

$$+ (CH_3)_2NH_2{}^+Cl^- + 1.5\,OP(C_6H_5)_3$$

In this reaction, triphenylphosphine serves both as a ligand and as a reductant. The CO ligand is derived from decomposition of some of the dimethylformamide (DMF) molecules.

One of the most striking chemical properties of Vaska's complex is its ability to bind O_2 reversibly:

$$IrCl(CO)[P(C_6H_5)_3]_2 + O_2 \rightleftharpoons IrCl(CO)[P(C_6H_5)_3]_2(O_2)$$

The oxygenation and deoxygenation reactions can be carried out simply by purging solutions of $IrCl(CO)[P(C_6H_5)_3]_2$ or $IrCl(CO)[P(C_6H_5)_3]_2(O_2)$ with O_2 or an inert gas, respectively. To some extent, the complex simulates the behavior of biological O_2 carriers such as hemoglobin, although the nature of the $M-O_2$ interaction in these cases is quite different. The binding of O_2 to Vaska's complex can be described as an oxidative addition in which the Ir^I center is converted to Ir^{III}. The bonding in this O_2 complex can be analyzed by infrared (IR) spectroscopy. In the second part of this experiment, you will study the reaction of Vaska's complex with O_2 and monitor the reaction by means of IR spectroscopy.

Infrared Spectra of Solids

Infrared spectroscopy is a powerful method for establishing the structure of a complex and for studying chemical bonding. This technique provides information about the vibrational modes of a molecule. Thus, the IR spectrum of Vaska's complex exhibits absorptions at frequencies (commonly expressed in wavenumbers (cm^{-1}), that is, reciprocal wavelength) characteristic of stretching and bending modes of the CO and $P(C_6H_5)_3$ groups. Although the Ir–Cl, Ir–P, and Ir–C stretching modes are also measurable, they generally occur at frequencies too low (<650 cm^{-1}) to be observed in most IR spectrophotometers. Commonly available IR spectrophotometers scan the range from 650 to 4000 cm^{-1}.

There are several ways to obtain an IR spectrum of a compound: (a) as the neat (undiluted) substance, (b) as a gas, for volatile compounds (see Experiment 6), (c) in solution, using a suitable solvent (see Experiment 17), (d) as a mull, or (e) as a compressed solid pellet. Which method is best depends on the nature of the compound under examination. In most of these methods, the sample is held between IR transparent windows. A potentially serious problem with windows is that they can become contaminated. If this occurs, then the windows will exhibit IR absorption bands due to the contaminants, and this is especially true with metal carbonyl impurities because their IR bands are intense. It is therefore important to clean IR sample cells properly.

For $IrCl(CO)[P(C_6H_5)_3]_2$, methods (a) and (b) are not practical, but the other methods are useful. The first of these other methods is solution IR spectroscopy (see Experiment 17). The other two IR methods involve solid samples of the compound of interest. Solid samples can be examined either as a mull or as a compressed pellet. In both these methods, the spectra will be affected by intermolecular interactions in the crystal. The intermolecular interactions sometimes produce changes in frequencies and even in the number of absorptions, compared with the solution spectra.

A *mull* is simply a slurry of the solid in a viscous inert liquid. The most commonly used mulling agent is Nujol, a brand name for purified mineral oil that consists of C_{20}–C_{30} alkanes. Occasionally, hexachlorobutadiene, C_4Cl_6, is used as a mulling agent if one is interested in analyzing the C–H bands in the sample. A drawback of hexachlorobutadiene is that it reacts with many organometallic compounds. The preparation of mulls that produce good spectra is an art. It is entirely normal to prepare several mulls before one obtains a satisfactory spectrum. Before interpreting the spectrum obtained from a mull, it is important to identify those absorptions that result from the mulling agent. The remaining absorption bands may then be assigned to the various vibrations of the complex. To make these assignments, it will be necessary to consult the references at the end of the experiment.

In the pellet technique, the desired compound is thoroughly mixed with an IR-transparent substance such as KBr. The resulting fine powder is compressed into a thin pellet with a diameter of approximately 1 cm. This sample preparation technique has the advantage that KBr has no absorptions of its own, although H_2O is a frequent contaminant. Sometimes one encounters problems because KBr reacts with the compound. If you examine the spectrum of $IrCl(CO)[P(C_6H_5)_3]_2$ as a KBr pellet, you should consider the possibility that Br^-/Cl^- exchange occurs to form $IrBr(CO)[P(C_6H_5)_3]_2$.

The most prominent absorption in the IR spectrum of Vaska's complex is the carbonyl (CO) stretching band, which is usually symbolized ν_{CO}. The vibrational frequency of CO ligands is sensitive to the electronic properties of the metal center by virtue of the overlap of the CO π^* orbital with metal d orbitals:

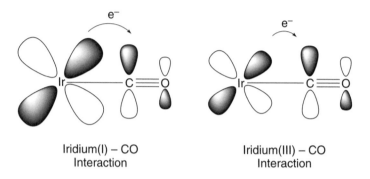

Iridium(I) – CO
Interaction

Iridium(III) – CO
Interaction

Generally, the transfer of electron density from the metal to the CO π^* orbital is more efficient when the metal is in a low oxidation state. This transfer of electron density, called π back-bonding, strengthens the M–C bond and weakens the C–O bond because the π^* orbital is antibonding with respect to C and O. This logic suggests that the ν_{CO} frequency in CO complexes will decrease as more π back-bonding occurs. These concepts are important for analyzing the bonding between metals and CO; the bonding of many other ligands, including alkenes, acetylenes, N_2, and isocyanides, can be analyzed in a similar way.

The characterization of CO complexes by IR spectroscopy is very convenient because the CO stretching bands occur in a region where few organic molecules absorb. Furthermore, CO stretching bands are usually rather intense; consequently, it is possible to record spectra from extremely dilute solutions. The ν_{CO} bands of metal carbonyls are even more intense than the ν_{CO} bands seen for organic carbonyls such as ketones and esters.

The ν_{CO} band for the iridium(I) compound $IrCl(CO)[P(C_6H_5)_3]_2$ is found near 1960 cm^{-1}, whereas in the iridium(III) compound $IrCl_3(CO)[P(C_6H_5)_3]_2$ it appears at 2120 cm^{-1}. This frequency shift indicates that the C–O bond is stronger (and the Ir–C bond weaker) in the latter complex. For the O_2 complex $IrCl(CO)[P(C_6H_5)_3]_2(O_2)$ the CO stretching band occurs at 2010 cm^{-1}, which suggests that the O_2 ligand withdraws electron density less effectively than two chloride ligands. In the present experiment, the relative amounts of $IrCl(CO)[P(C_6H_5)_3]_2$ and $IrCl(CO)[P(C_6H_5)_3]_2(O_2)$ can be estimated from the relative intensities of the ν_{CO} bands at 1960 and 2010 cm^{-1}.

EXPERIMENTAL PROCEDURE

Note: Iridium salts can cost up to US$100 per gram, so it is essential to use small quantities in this experiment. When small amounts of compounds are being handled, it is important to minimize mechanical losses by using small reaction vessels and by reducing the number of times the sample is transferred.

trans-IrCl(CO)[P(C$_6$H$_5$)$_3$]$_2$

In a 50-mL round-bottom flask mounted in a rheostat-controlled heating mantle, place a Teflon-coated stir bar, 0.105 g (0.3 mmol) of IrCl$_3 \cdot$3 H$_2$O,* 0.393 g (1.5 mmol) of triphenylphosphine, 10 mL of dimethylformamide, and 0.120 mL of aniline. Fit the flask with a reflux condenser topped with a N$_2$ inlet connected to a bubbler. Briefly flush the air from the system with a N$_2$ stream. Decrease the flow of N$_2$ gas and heat the mixture to vigorous reflux for 2 h (or longer at higher altitudes). Heating can be continued for up to 24 h without adversely affecting the yield. Allow the yellow solution to cool to 50 °C (or below) and then add 30 mL of methanol. Cool the mixture in an ice bath for 10 min and then collect the yellow microcrystals by filtration in air. Wash the crystals with a few milliliters each of methanol and diethyl ether, and then dry the solid in a vacuum. Solid samples of Vaska's complex are air stable. Record the yield, and save some of the product for an IR spectrum.

Oxygenation of IrCl(CO)[P(C$_6$H$_5$)$_3$]$_2$

In a 100-mL Schlenk flask containing a magnetic stirring bar, prepare a solution of 40 mg of Vaska's complex in 20 mL of toluene (see Experiment 17 for a discussion of Schlenkware). While stirring the solution, flush the flask with O$_2$ gas, then close off the stopcock and stopper the flask. Vigorously stir the solution in the O$_2$ atmosphere for 90 min. Evaporate the solution to obtain a yellow-orange residue. Record the IR spectrum of some of the solid as a Nujol mull. Time permitting, redissolve the remainder in toluene and purge the solution again with a stream of N$_2$ gas. Once the solvent has evaporated, remove some of the solid residue and examine its IR spectrum.

IR Spectra as Nujol Mulls

This section describes in detail a method for preparing mulls of compounds.

1. Put about 5 mg of the complex in a polished agate mortar and grind the sample with an agate pestle until the substance forms a glossy layer in the mortar (between 1 and 5 min of grinding). The extent of grinding will make the difference between a good and a bad spectrum.
2. Add a small drop of the Nujol (or other mulling agent) to the lip of the mortar. With the pestle, transfer a *small* amount of the mulling agent to the powder. Grind this mixture, adding more mulling agent if necessary, until all of the sample has been converted into a paste.

---

*This procedure can be conducted using an equivalent molar amount of other iridium salts, such as H$_2$IrCl$_6$. The synthesis tolerates small amounts (0.5 mL) of water.

3. With a rubber policeman or spatula, transfer this paste to a single-crystal NaCl disk (often called a "salt plate"). The preparation of mulls may scratch the plates, but this will not normally affect the spectra.
4. Place a second NaCl plate on top of the paste and gently rotate the plates until the paste spreads to a thin film between the plates.
5. Put the plates in a holder such as the one shown in Figure 19-1, and scan the IR spectrum of the sample. See your instructor for the operation of your

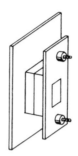

Figure 19-1
Plate holder for Nujol mulls.

particular spectrophotometer. If you are using a grating instrument, leave the reference beam of the instrument open. No reliable compensation for the mulling agent can be made by inserting a sample of it in the reference beam. If you are using a Fourier transform (FT) IR instrument, collect a background spectrum before putting the sample in the beam.
6. Clean the NaCl plates by rinsing them with dry acetone and, if necessary, by wiping them with a soft cloth dampened with acetone. After drying the plates in air, return them to a desiccator.

REPORT

Include the following:
1. Yield of $IrCl(CO)[P(C_6H_5)_3]_2$.
2. IR spectra of $IrCl(CO)[P(C_6H_5)_3]_2$ and $IrCl(CO)[P(C_6H_5)_3]_2(O_2)$.
3. Relative intensities of the ν_{CO} peaks for $IrCl(CO)[P(C_6H_5)_3]_2$ and $IrCl(CO)[P(C_6H_5)_3]_2(O_2)$ after oxygenation and after the nitrogen purge. (*Note*: If your IR spectrometer gives the intensities in transmittance units, you will have to convert the values to absorbance units before computing the ratio of intensities.)

PROBLEMS

1. How does the ν_{CO} stretching frequency of coordinated CO compare with free CO (2310 cm^{-1}), and how can the difference be explained in terms of bonding (σ and π) with the metal?
2. What is the importance of oxidative addition reactions in catalysis?
3. How much does it cost to purchase $IrCl_3 \cdot 3\,H_2O$ and Vaska's complex? Evaluate the total cost of the materials used in your synthesis.
4. What is the role of the aniline in the synthesis of Vaska's complex?

INDEPENDENT STUDIES

A. Treat $IrCl(CO)[P(C_6H_5)_3]_2$ with dppe, $(C_6H_5)_2PCH_2CH_2P(C_6H_5)_2$, (prepared in Experiment 8) to obtain $[Ir(dppe)_2]Cl$. Examine the reactivity of this complex with O_2. (Vaska, L.; Catone, D. L. *J. Am. Chem. Soc.* **1966**, *88*, 5324.)

B. Examine the reaction of $IrCl(CO)[P(C_6H_5)_3]_2$ with HCl to give $IrHCl_2(CO)[P(C_6H_5)_3]_2$. (Blake, D. M.; Kubota, M. *Inorg. Chem.* **1970**, *9*, 989.)

C. Prepare the iodide $IrI(CO)[P(C_6H_5)_3]_2$ and examine its oxygenation. (McGinnety, J. A.; Doedens, R. J.; Ibers, J. A. *Inorg. Chem.* **1967**, *6*, 2243.)

D. Convert Vaska's complex into the dinitrogen compound $IrCl(N_2)[P(C_6H_5)_3]_2$. (Collman, J. P.; Hoffman, N. W.; Hosking, J. W. *Inorg. Synth.* **1970**, *12, 8*.)

E. Prepare the 1,5-cyclooctadiene (cod) complex $Ir_2Cl_2(cod)_2$. (Herdé, J. L.; Lambert, J. C.; Senoff, C. V. *Inorg. Synth.* **1982**, *15*, 18.)

F. Prepare the 1,5-cyclooctadiene (cod) complex $\{Rh(cod)[P(C_6H_5)_3]_2\}PF_6$ and study its ligand substitution reactions. (Kruger, H.; de Waal, D. J. A. *J. Chem. Educ.* **1987**, *64*, 262.)

G. Use microscale techniques to prepare $RhCl(CO)[P(C_6H_5)_3]_2$. (Singh, M. M.; Szafran, Z.; Pike, R. M. *J. Chem. Educ.* **1990**, *67*, A180.)

H. Prepare the 1,5-cyclooctadiene complex $PdCl_2(cod)$ and study its reaction with sodium methoxide. (Bailey, C. T.; Lisensky, G. C. *J. Chem. Educ.* **1985**, *62*, 897.)

REFERENCES

Vaska's Complex

Barton, D. H. R.; Martell, A. E.; Sawyer, D. T. Eds. *The Activation of Dioxygen and Homogeneous Catalytic Oxidation*, Plenum: New York, 1993.

Collman, J. P.; Hegedus, L. S.; Norton, J. R.; Finke, R. G. *Principles and Application of Organotransition Metal Chemistry*, University Science Press: Mill Valley, CA, 1987; Chapter 5.

Cotton, S. A. *Chemistry of Precious Metals*, Blackie: London, 1997.

Vaska, L.; DiLuzio, J. W. *J. Am. Chem. Soc.* **1961**, *83*, 2784. The original paper on Vaska's complex.

Vrieze, K.; Collman, J. P.; Sears, C. T. Jr.; Kubota, M. *Inorg. Synth.* **1968**, *11*, 101. Preparation of Vaska's complex.

Microscale Techniques

Shriver, D. F.; Drezdzon, M. A. *The Manipulation of Air Sensitive Compounds*, 2nd ed., Wiley: New York, 1986.

Szafran, Z.; Pike, R. M.; Singh, M. M. *Microscale Inorganic Chemistry*, Wiley: New York, 1991.

Infrared Spectroscopy

Ferraro, J . R. *Low-Frequency Vibrations of Inorganic and Coordination Compounds*, Plenum: New York, 1971. Excellent source of practical and theoretical information in this area.

Nakamoto, K. *Infrared Spectra of Inorganic and Coordination Compounds*, 5th ed., Wiley: New York, 1997. Very useful to an inorganic chemist.

Nyquist, R. A.; Kagel, R. O. *Infrared Spectra of Inorganic Compounds*, Academic: New York, 1971. Figures of spectra; no organometallic compounds.

The Air-Sensitive Sandwich Complex Nickelocene

Note: This experiment requires about 8 hours for Part A and 4 hours for Part B.

The cyclopentadienyl ligand C_5H_5 (abbreviated Cp) occupies an important position in organometallic chemistry, the field that deals with the chemistry of metal–carbon bonds. In fact, the current widespread interest in organometallic chemistry can be traced to the discovery in 1952 that dicyclopentadienyl iron, $Fe(C_5H_5)_2$, has a structure in which each of the cyclopentadienyl groups is bound to the iron center by means of all five carbon atoms:

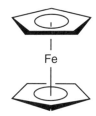

This discovery was made almost simultaneously by two people: the English chemist Geoffrey Wilkinson and the German chemist Ernst Otto Fischer. Before their work, dicyclopentadienyl iron was thought to have a structure in which each of the the Cp rings was bound to the iron center by only one iron–carbon bond. Such a structure could not, however, account for the incredible stability of $Fe(C_5H_5)_2$: It is stable to 470 °C and is unaffected by boiling concentrated HCl. Wilkinson and Fischer's achievement lay in the recognition that the unusual chemical properties of the molecule had to reflect an unusual structure. The type of bonding between the metal and the cyclopentadienyl group is now known as π bonding because the metal interacts with the π orbitals on the ligand. Wilkinson and Fischer showed that such bonds could be formed with many other metals and with many other ligands. For this and for their later work, they shared the Nobel prize in 1973.

The robust nature of the metal–cyclopentadienyl bond is one of the reasons that organometallic compounds have been so extensively investigated since 1952. Cyclopentadienyl complexes have many uses. For example, they are used as components of catalysts for the manufacture of polyethylene and polypropylene. In some parts of the world, in fact, the methylcyclopentadienyl compound $(C_5H_4CH_3)Mn(CO)_3$ is used as an anti-knock additive in gasoline.

Dicyclopentadienyl iron is only one representative of a whole family of organometallic compounds. Complexes of stoichiometry $M(C_5H_5)_2$ are also known for vanadium, chromium, manganese, ruthenium, osmium, cobalt, and nickel. Collectively, these $M(C_5H_5)_2$ compounds are known as the metallocenes; this name was coined to reflect the aromatic (arene) character of the metal-bound Cp rings. The metallocenes are also known as "sandwich" complexes because the two cyclopentadienyl rings can be thought of as two pieces of bread, with the central metal serving as the filling.

Cyclopentadiene is prepared by the thermal cracking (dedimerization) of its Diels–Alder dimer, dicyclopentadiene, whose IUPAC name is 3a,4,7,7a-tetrahydro-4,7-methanoindene. The dimer undergoes cracking when it is heated above 100 °C:

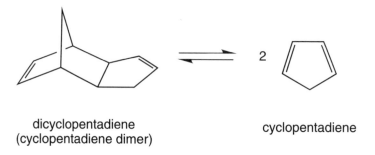

dicyclopentadiene
(cyclopentadiene dimer)

cyclopentadiene

Once prepared by this route, cyclopentadiene is stable for several hours at room temperature but eventually reverts back to the dimer. Thus, it should either be used immediately after it is prepared or be stored in a flask cooled in a dry ice bath.

There are several ways to prepare Cp compounds from cyclopentadiene. By far, the most popular method involves sodium cyclopentadienide, NaC_5H_5, which can be prepared from cyclopentadiene by treating it with sodium hydride or sodium metal. Sodium cyclopentadienide, when rigorously pure, forms a nearly colorless solution in tetrahydrofuran (THF), but it reacts readily with oxygen to form highly colored violet and brown materials. Under the best conditions typically used, solutions of sodium cyclopentadienide are pink. The color of the solution is an indication of the skill of the researcher: the lighter the color, the better the technique. In practice, however, the yields of cyclopentadienyl derivatives appear to be largely unaffected by the color of the sodium cyclopentadienide solutions, unless the latter are so extensively oxidized as to be dark brown.

In Part A of this experiment, the sodium cyclopentadienide is used to make nickelocene, $Ni(C_5H_5)_2$. Nickelocene is an interesting organometallic compound because it has a valence electron count of 20: This number exceeds the "normal" 18-electron count seen for many organometallic compounds. The extra two electrons are in orbitals that are weakly metal–carbon antibonding; as a result, nickelocene often undergoes reactions in which metal–carbon bonds are broken and the electron count at the metal center changes to 18. Nickelocene forms dark green crystals that oxidize slowly in air over a period of several days. Crystals of nickelocene can thus be handled for brief periods in air, but powdered samples decompose very quickly in air. Thus, the handling and storing of nickelocene are usually carried out under an inert atmosphere. The reactions used to prepare nickelocene are

$$[Ni(H_2O)_6]Cl_2 + 6\,NH_3 \longrightarrow [Ni(NH_3)_6]Cl_2 + 6\,H_2O$$

$$C_5H_6 + NaH \longrightarrow NaC_5H_5 + H_2$$

$$2\,NaC_5H_5 + [Ni(NH_3)_6]Cl_2 \longrightarrow Ni(C_5H_5)_2 + 2\,NaCl + 6\,NH_3$$

In the second part of this experiment, nickelocene is treated with dimethyl acetylenedicarboxylate (DMAD). The acetylene ligand adds to one of the Cp rings in a Diels–Alder reaction to give a bicyclo[2.2.1]heptadienyl unit coordinated to the nickel center. True to form, in this reaction, two nickel–carbon bonds are broken and the electron count of the product is 18:

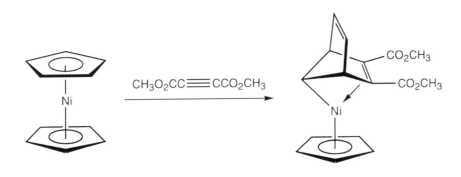

EXPERIMENTAL PROCEDURE

Part A

Cyclopentadiene, C_5H_6

Safety note: Dicyclopentadiene and cyclopentadiene are toxic and have unpleasant musty odors. They should be handled in a hood.

Freshly prepared cyclopentadiene is needed for this experiment, and it is best to set up this apparatus first because the distillation step takes some time. In a hood, place 50 mL of dicyclopentadiene into a 500-mL two-neck round-bottom flask. A large flask should be used because heating the cyclopentadiene dimer tends to give large amounts of foam, which can bubble over during the distillation step and contaminate the product. Equip the round-bottom flask with with a short (15 cm) Vigreux column topped by a distillation head with a thermometer adapter. The distillation head should be attached to a water condenser and a vacuum adapter (the latter should have a drip tube); finally, the vacuum adapter should be connected to a two-neck 100-mL receiving flask. Attach and secure water hoses to the condenser and turn on the water. Plug the second neck

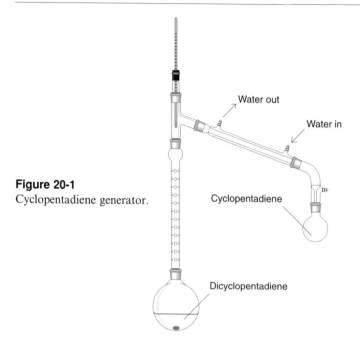

Figure 20-1
Cyclopentadiene generator.

of the distillation flask with a glass stopper and plug the second neck of the receiving flask with a septum. Use silicone grease on all glass-to-glass joints. The apparatus is pictured in Figure 20-1.

Attach a hose from a N_2 manifold to the vacuum adapter, and purge the apparatus with N_2 for about 2 min. Heat the distillation flask to about 130 °C. At this temperature, which is below the boiling point of dicyclopentadiene, the liquid will begin to foam as cyclopentadiene monomer is generated. The vapor of the monomer travels up the Vigreux column, and the liquid that distills into the receiver should have a boiling point of about 40 °C as measured by the thermometer at the top of the distillation head. Do not allow the temperature in the distillation head to rise above 40 °C, because this would indicate that some of the dimer has reached the top of the column without cracking. If this happens, reduce the temperature of the heating bath.

Continue the distillation until 15–20 mL of cyclopentadiene monomer has been collected, and then stop the heating. While the dicyclopentadiene is being cracked, carry out the following synthesis of $[Ni(NH_3)_6]Cl_2$.

If other students need freshly prepared cyclopentadiene, then the distillation apparatus may be left in place for reuse (ask your instructor). If you leave the apparatus in place, pour any unused distillate back into the distillation flask at the end of the laboratory period.

Hexamminenickel(II) Chloride, $[Ni(NH_3)_6]Cl_2$

Safety note: Ammonium hydroxide is corrosive and has a pungent penetrating odor. Carry out all operations in a hood.

In a 250-mL beaker, dissolve 8 g (0.034 mol) of nickel(II) chloride hexahydrate, $NiCl_2 \cdot 6 H_2O$, in 20 mL of distilled H_2O. If necessary, filter the solution

through a glass frit to remove impurities. To this solution, slowly but continuously add 30 mL of concentrated ammonium hydroxide, NH_4OH. Cool the
resulting solution by placing the beaker in an ice bath, and then add 80 mL of
ethanol to complete the precipitation of $[Ni(NH_3)_6]Cl_2$. Collect the lavender solid
by filtration and wash it with 20 mL of ethanol and 30 mL of diethyl ether. Allow
the solid to air dry. Record the yield.

Dicyclopentadienylnickel(II), $Ni(C_5H_5)_2$

Safety note: Sodium hydride reacts violently with water and the hydrogen
evolved can explode. Dispose of sodium hydride by treating it with isopropanol.

Place a magnetic stir bar into a 200-mL three-neck flask fitted with a gas
inlet, a condenser, and a rubber septum (see Fig. 20-2). The condenser should
be attached to a source of N_2 gas and an oil bubbler. Remove the stopper,
and flush the flask with N_2 for 1 or 2 min. While flushing with N_2, add to the
flask 1.0 g (0.022 mol) of a 60% by weight dispersion of sodium hydride in
mineral oil. Add 40 mL of hexane to the flask, and stir briefly to dissolve the oil.
Stop the stirring and let the sodium hydride particles settle to the bottom of the
flask for a few minutes. Carefully remove most of the hexane layer with a large
syringe or pipet. Discard the hexane wash into a 250-mL beaker containing 50
mL of 95% ethanol. (If the sodium hydride particles do not settle well, remove as

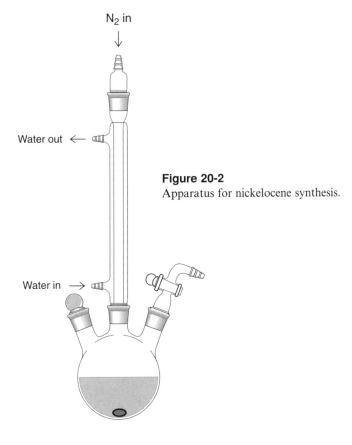

N₂ in

Water out ←

Figure 20-2
Apparatus for nickelocene synthesis.

Water in →

much of the particle-free hexane layer as is possible, add another 40 mL of hexane, stir briefly, and let the contents of the flask settle for 5–10 min. Then use a syringe or pipet to remove the upper hexane layer. It is all right to leave 10 mL or so of hexane in the flask.)

Add 40 mL of freshly distilled tetrahydrofuran (THF) to the flask and then plug the joint with a septum. With a syringe, slowly add 7 mL (0.087 mol) of freshly cracked cyclopentadiene to the NaH–THF suspension. There should be an exothermic reaction and evolution of gas. If unreacted sodium hydride remains, the solution should be warmed in a water bath; if some sodium hydride still remains, add some additional cyclopentadiene (1 mL). After the sodium hydride has reacted, cool the solution in an ice bath (as you do this, watch the bubbler to prevent developing a negative pressure). Against a N_2 flow, carefully add 3.5 g (0.015 mol) of $[Ni(NH_3)_6]Cl_2$ in approximately 1-g portions every 5 min. Remove the septum for these additions. After the last addition of the nickel compound, replace the septum with a greased glass stopper. Slowly warm the solution until the reaction begins to occur, which is signaled by the evolution of ammonia gas. (*Caution*: If the reaction becomes too vigorous, stop the stirring and cool the flask briefly in ice.) Stir the warm solution for an hour, then add 1 mL of ethanol to destroy any unreacted NaC_5H_5. Replace the condenser with a greased glass stopper, and attach the flask to a Schlenk line (see Fig. 17-2). After you make sure that the Dewar surrounding the Schlenk line's trap is filled with liquid N_2, remove the solvent under vacuum. After the solvent has been removed, fill the flask with N_2, close the stopcock, and secure all stoppers. *Here is a good stopping point for the first day's work.* The sublimation will be done during the second week.

Purification of Nickelocene

The crude solid from the procedure above will be purified by sublimation. Obtain a cold finger that has a ground glass joint of the proper size to fit into the joint of your flask (Fig. 20-3). The cold finger should be long enough to reach within

Figure 20-3
A cold finger sublimer. (Courtesy of Kimble/Kontes Glass Inc.)

about 2 cm of the bottom of the flask. Attach the flask containing the crude nickelocene to a Schlenk line that has both a source of N_2 gas and a vacuum (see Fig. 17-2). With the stopcock on the flask still closed, evacuate the hose connecting the flask to the line for 1 min. Then fill the hose with N_2 and open the stopcock on the flask. Remove the glass stopper. Use a "magnet wand" to remove the stir bar from the flask. Then insert the cold finger, using silicone grease to lubricate the joint. Close the stopcock on the flask, evacuate the hose, and then slowly reopen the stopcock to evacuate the flask. Attach hoses to the cold finger and run cold water through it. Sublime the product by warming the lower portion of the flask to about 80 °C using an oil bath. Do *not* exceed 120 °C during this step or decomposition products will contaminate the sublimate. The progress of the sublimation is indicated by the appearance of green crystals on the cold finger. After the flask has been kept at 80 °C for about 1 h, cool the flask to room temperature, and fill the flask with N_2. Close the stopcock on the flask. Detach the water cooling hoses, and carefully invert the flask to let the water drain out of the cold finger. Cover the water entrance and exit ports on the cold finger with some Parafilm®.

The next step, which involves collecting the sublimed nickelocene, is best conducted in a glovebag. The use of a glovebag requires planning of experimental operations before they are actually carried out in the bag. It is convenient to stabilize the bag by putting a couple of ring stand bases inside the bag before introducing your equipment.

The following equipment is needed for collecting the sublimed nickelocene.

> A spatula
> A wide-mouth funnel for transferring solids
> A plastic weighing boat
> A Schlenk flask with stopcock sidearm.
> A rubber septum for the joint of the Schlenk flask
> The stoppered flask containing the sublimed nickelocene

Before putting the equipment into the bag, grease a stopcock key and place it in the stopcock on the Schlenk flask. Put the rubber septum in the neck and weigh the flask and septum together. This weighing will eventually be used to determine the amount of nickelocene that you have sublimed. Also, double check that most of the water has been drained from the cold finger and that the entrance ports on the cold finger are sealed with Parafilm.

Connect the glovebag (Fig. 20-4) to a N_2 cylinder and allow the gas to flush the bag for about 5 min; then place the equipment above into the bag. With the entrance port of the bag still open and the N_2 flowing, push on the top of the bag to make it as small as possible and then roll up the flaps at the entrance port and clamp them closed. When the N_2 has filled the bag to a convenient size (it should be flabby and not completely full), turn off the N_2 flow. Powder your hands with talcum powder or wear cotton gloves before putting your hands into the gloves in the bag. All operations are more difficult in the bag, and a spilled compound or solution makes a mess: take your time and work slowly.

Open the stopcock on the flask containing the sublimed nickelocene. Carefully withdraw the cold finger from the flask, trying not to bump the sublimed

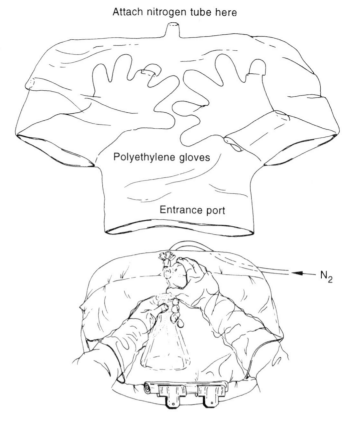

Figure 20-4
Glovebag before and after inflation.

nickelocene against the grease on the inside of the joint. Using the spatula, scrape the sublimed nickelocene into the plastic weighing boat, and from there into the second flask using the funnel. When done, plug the joint on the Schlenk flask with a septum, and turn the stopcock shut. Open the glovebag and remove all the equipment. Weigh the preweighed flask containing the transferred nickelocene. When done with the weighing, reattach the flask to the Schlenk line, and with the flask attached to a N_2 source and the stopcock open, replace the septum with a greased glass stopper. Use a joint clamp or rubber bands to secure the stopper. With the stopcock closed, the N_2-filled flask will protect the nickelocene. The product can be stored this way indefinitely. Pure nickelocene is crystalline and dark green. If your sublimed product is impure (has brown impurities or is oily), a second sublimation will be necessary. For long-term storage, the nickelocene should be placed into an ampule and flame-sealed. See your instructor.

Characterize your product by melting point and infrared (IR) spectroscopy (see Experiment 19 for the preparation of a Nujol mull). The determination of the melting point must be carried out under N_2. You can use a N_2-flushed glovebag to put some of your compound in a melting point capillary; plug the end of the capillary with grease to protect the nickelocene from air.

Part B

Reaction of Nickelocene with Dimethyl acetylenedicarboxylate
Safety note: The compound $CH_3O_2CC\equiv CCO_2CH_3$ is a lachrymator (tear gas!). Use it in a hood.

In a round-bottom flask, place 1 g (5.2 mmol) of $Ni(C_5H_5)_2$. Flush the flask with N_2 and place a septum in the neck. Add 0.7 g (5.8 mmol) of DMAD and 30 mL of THF to the flask with a syringe. Let the mixture stir for 1 week. Remove the solvent on a rotary evaporator, and then dissolve the product in 5 mL of chloroform. Purify the product by chromatography on a 15-cm tall, 3-cm diameter column of neutral alumina using a 1:1 solution of hexane and chloroform as the elutant (see Experiment 23 for more information about chromatography). Discard the colorless first fraction. Collect the orange band, and remove the solvent on a rotary evaporator. Dissolve the residue in 10–20 mL of hot hexane. Crystals should form immediately upon swirling the flask in an ice bath. If they do not, boil off some of the hexane, and then recool the solution in an ice bath. Collect the crystals by filtration. Record the yield. Characterize the product by melting point, IR (Nujol mull), and proton nuclear magnetic resonance (^{1}H NMR) spectroscopy (solvent: $CHCl_3$ or $CDCl_3$).

REPORT

Include the following:
1. Yields of $Ni(C_5H_5)_2$ and its Diels–Alder adduct with DMAD.
2. Melting points of $Ni(C_5H_5)_2$ and its Diels–Alder adduct with DMAD.
3. IR spectra of $Ni(C_5H_5)_2$ and its DMAD adduct. Point out similarities and differences.
4. ^{1}H NMR spectrum of the DMAD adduct and assignment of the peaks.

PROBLEMS

1. Why is nickelocene of interest in terms of the 18-electron rule, and how does its electron count relate to its reactivity? What is the electron count of $Ni(NH_3)_6^{2+}$?
2. Is nickelocene diamagnetic or paramagnetic? What effect will its magnetism have on its ^{1}H NMR spectrum?
3. Why do we use DMAD rather than some other alkyne? Why is this alkyne sometimes described as "activated"?
4. Of the two compounds $Fe(C_5H_5)_2$ and $Ni(C_5H_5)_2$, which is easier to oxidize and why?
5. What features make a compound sublime easily?

INDEPENDENT STUDIES

A. Obtain the ^{1}H NMR spectrum of Ni(C$_5$H$_5$)$_2$. (Köhler, F. H. *J. Organomet. Chem.* **1976**, *110*, 235.)

B. Record the mass spectra of Ni(C$_5$H$_5$)$_2$ and its DMAD adduct and assign the peaks.

C. Carry out the reaction of Ni(C$_5$H$_5$)$_2$ with a different alkyne. (Brunner, H.; Pieronczyk, W. *Bull. Soc. Chim. Belg.* **1977**, *86*, 725.)

D. Prepare nickelocene using KOH and C$_5$H$_6$ to make KC$_5$H$_5$. (Barnett, K. W. *J. Chem. Educ.* **1974**, *51*, 422.)

E. Prepare bis(indenyl)iron. (Westcott, S.; Kakkar, A. K.; Stringer, G.; Taylor, N. J.; Marder, T. B. *J. Organomet. Chem.* **1990**, *394*, 777.)

F. Carry out the reaction of ferrocene with acetyl chloride to form Fe(C$_5$H$_5$)(C$_5$H$_4$COCH$_3$). (Bozak, R. E. *J. Chem. Educ.* **1966**, *43*, 73; Graham, R. J.; Lindsey, R. V.; Parshall, G. W.; Peterson, M. I.; Whitman, G. M. *J. Am. Chem. Soc.* **1957**, *79*, 3416.)

G. Prepare (C$_5$H$_5$)$_2$TiS$_5$ and study its conformational dynamics by NMR spectroscopy. (Diaz, A.; Radzewich, C.; Wicholas, M. *J. Chem. Educ.* **1995**, *72*, 937.)

REFERENCES

Nickelocene

Barnett, K. W. *J. Chem. Educ.* **1974**, *51*, 422. Preparation of nickelocene.

Bleasdale, C.; Jones, D. W. *J. Chem. Soc., Perkin Trans. 1* **1986**, 157. Use of the nickelocene–DMAD adduct in organic synthesis.

Dahl, L. F.; Wei, C. H. *Inorg. Chem.* **1963**, *2*, 713. Structure of the nickelocene–DMAD adduct.

Dubeck, M. *J. Am. Chem. Soc.* **1960**, *82*, 6193. Reaction of nickelocene with DMAD.

Fischer, E. O.; Jira, R. *Z. Naturforsch. B* **1953**, *8*, 217. Original synthesis of nickelocene.

Köhler, F. H. *J. Organomet. Chem.* **1976**, *110*, 235. The ^{1}H and ^{13}C NMR spectra of metallocenes.

Seiler, P.; Dunitz, J. D. *Acta Crystallogr., Sect. B* **1980**, *36*, 2255. Structure of nickelocene.

Wilkinson, G.; Pauson, P. L.; Birmingham, J. M.; Cotton, F. A. *J. Am. Chem. Soc.* **1953**, *75*, 1011. Original synthesis of nickelocene.

General References to Cyclopentadienyl Complexes

Abel, E. W.; Stone, F. G. A.; Wilkinson, G., Eds., *Comprehensive Organometallic Chemistry II*, Pergamon: New York, 1995. Contains many chapters on metallocene complexes.

Bard, A. J.; Garcia, E.; Kukharenko, S.; Strelets, V. V. *Inorg. Chem.* **1993**, *32*, 3528. Electrochemistry of metallocenes.

Bublitz, D. E.; Rinehart, K. L. *Organic Reactions* **1969**, *17*, 1. The synthesis of substituted ferrocenes and other cyclopentadienyl transition metal complexes.

Holloway, J. D. L.; Geiger, W. E. *J. Am. Chem. Soc.* **1979**, *101*, 2038. Redox properties of metallocenes.

Kimel'feld, Y. M.; Smirnova, E. M.; Aleksanyan. V. T. *J. Mol. Struct.* **1973**, *19*, 329. The IR spectra of metallocenes.

King, R. B. *Organometallic Syntheses*, Academic: New York, 1965, p 64. Syntheses of bis(cyclopentadienyl)metal derivatives.

Togni, A.; Hayashi, T., Eds., *Metallocenes*, Wiley-VCH: Weinheim, 1998. Review of the field.

Wilkinson, G. *J. Organomet. Chem.* **1975**, *100*, 273. A personal account of the discovery of the structure of ferrocene.

Glove Box and Glove Bag Techniques

Shriver, D. F.; Drezdzon, M. A. *The Manipulation of Air-sensitive Compounds*, 2nd ed., Wiley: New York, 1986. Various gloveboxes and glovebags and their use.

Part V

BIOINORGANIC CHEMISTRY

Cobaloximes: Models of Vitamin B$_{12}$

Note: This experiment requires one 3- or 4-hour laboratory period.

Cobalt is a trace element in living systems, and in humans is present at a concentration of only 10^{-8} M. This low concentration does not, however, indicate that cobalt is biologically unimportant: A deficiency of cobalt in the human diet leads to the fatal disorder pernicious anemia. In the body, cobalt is present in the family of coenzymes called cobalamins, one derivative of which is cyanocobalamin or vitamin B$_{12}$. Coenzymes are compounds that bind to proteins to form enzymes, nature's catalysts. As the first organometallic complexes found in nature, the cobalt-containing coenzymes have attracted much attention from both inorganic and biological chemists.

The structures of all cobalamins are similar and consist of a cobalt ion situated in a tetrazamacrocyclic ligand with trans imidazole (symbolized by L) and alkyl groups. The structure of methylcobalamin, in which the alkyl group is a methyl group, is shown in Figure 21-1.

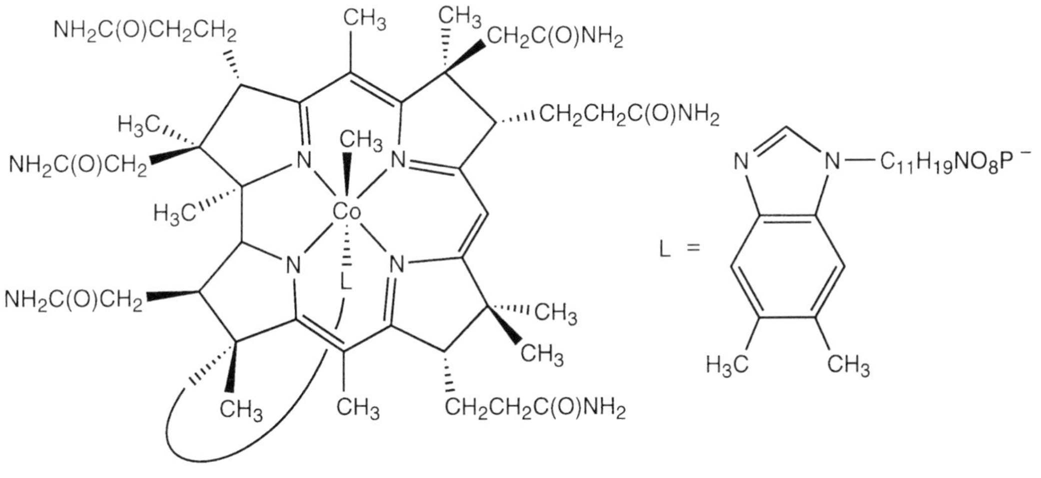

Figure 21-1
Structure of methyl cobalamin.

At various stages in the enzyme's catalytic cycle, the oxidation state of cobalt can be Co^{III}, Co^{II}, or Co^{I}. The macrocyclic ligand, called a corrin, resembles porphyrins (Experiment 23) except that the ring contains one less carbon atom and it is more reduced, consisting of only 13 contiguous sp^2 atoms. In the coenzyme, the Lewis base L is an N-substituted dimethylbenzimidazole. Recent evidence suggests that in the coenzyme-protein complex (holoenzyme) the Lewis base is sometimes replaced by the imidazole group of a histidine residue.

The most important component of the coenzyme is the cobalt–carbon bond, because this bond is cleaved during the enzyme's catalytic cycle. Cobalamin-based enzymes catalyze two kinds of transformations, depending on the identity of the alkyl ligand. When the alkyl group is methyl, the enzymes are involved in methylation reactions such as the biosynthesis of the amino acid methionine:

(homocysteine)   (methionine)

The occurrence of the poisonous substance methyl mercury in the environment has been attributed to the action of bacterial methylcobalamins on inorganic mercury salts. In the second class of cobalamin-based enzymes, in which the alkyl ligand is adenosyl, the enzyme catalyzes carbon skeletal rearrangements. For example, one adenosylcobalamin-based enzyme catalyzes the conversion of methylmalonate to succinate, a key step in the Krebs citric acid cycle:

(methylmalonate)   (succinate)

Unlike methylcobalamin (which serves as a source of CH_3^+), adenosylcobalamin initiates reactions by means of the reversible homolysis of the cobalt–adenosyl bond. The resulting carbon-based adenosyl radical abstracts a hydrogen atom from the substrate, thereby initiating the subsequent skeletal rearrangements.

The molecular complexity of the cobalamins makes it difficult to study the coordination chemistry of the cobalt center. Inorganic chemists, however, have found that many properties of these enzymes can be simulated using model compounds derived from fairly simple ligands. The most successful model complexes are known as the cobaloximes. In the cobaloximes, a pair of dimethylglyoximate ligands serves the function of the corrin macrocycle. Drawings of dimethylglyoxime and its conjugate base, the dimethylglyoximate anion, are shown below:

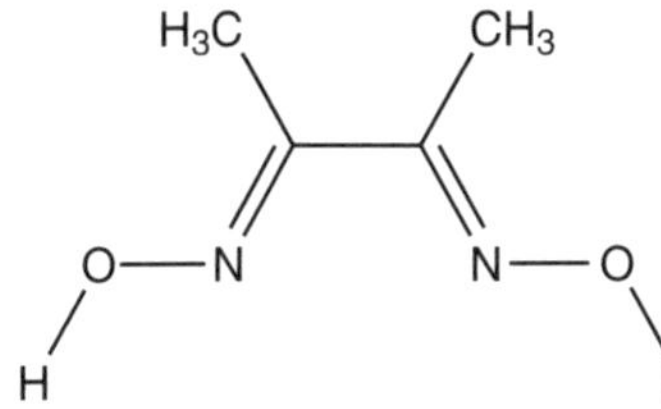

dimethylglyoxime
(dmgH$_2$)

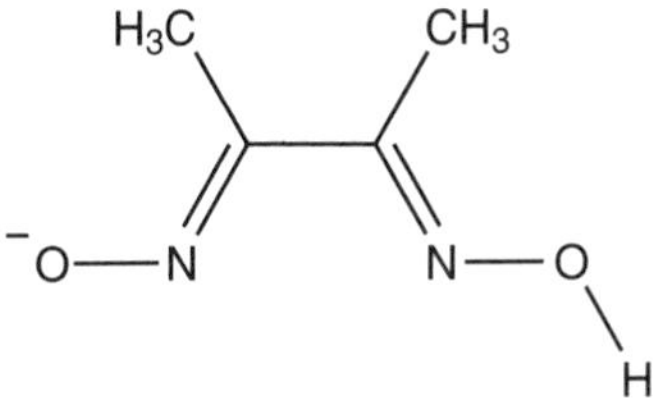

dimethylglyoximate
(dmgH$^-$)

Two dimethylglyoximate (dmgH$^-$) ligands can link through hydrogen bonds to form a planar anionic macrocycle that binds the cobalt through its four nitrogen atoms. An important feature of these compounds is that the nitrogen ligands are sp^2-hybridized, in contrast to the sp^3-hybridized amine ligands usually found in classical cobalt coordination compounds (see Experiments 13 and 14).

The present experiment begins with the preparation of Co(dmgH)$_2$(py)$_2$, where py = pyridine, although this cobalt(II) complex is not isolated. Reduction of this complex with BH$_4^-$ produces the cobalt(I) derivative Co(dmgH)$_2$(py)$^-$. This highly nucleophilic complex reacts with alkyl halides to give cobalt(III) alkyl complexes. In this experiment, you will use the alkylating agent isopropyl bromide (2-bromopropane):

$$Co(dmgH)_2(py)_2 + NaBH_4 \longrightarrow Na[Co(dmgH)_2(py)] + BH_3 \cdot py + \tfrac{1}{2} H_2$$

$$Na[Co(dmgH)_2(py)] + BrCH(CH_3)_2 \longrightarrow Co[CH(CH_3)_2](dmgH)_2(py) + NaBr$$

The structure of the compound you will make is shown below:

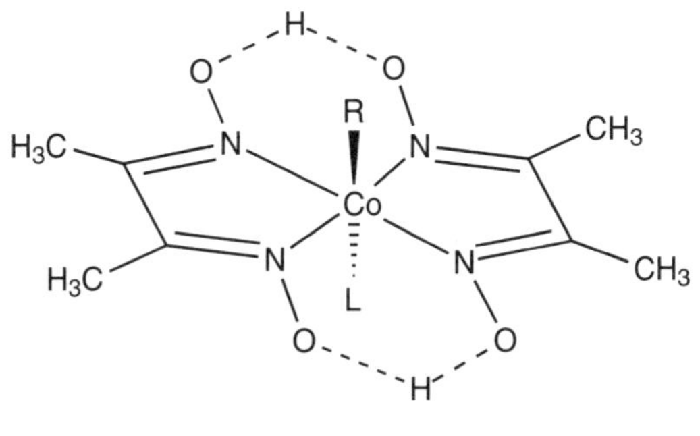

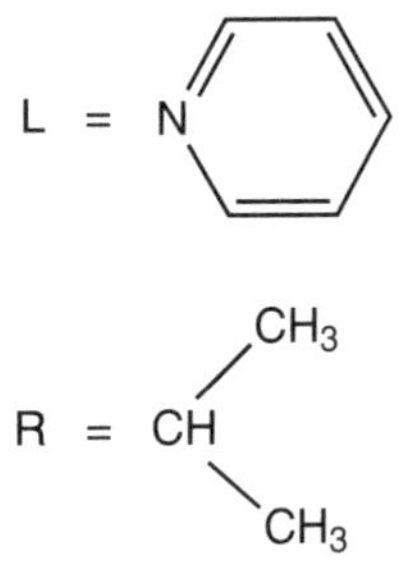

CoR(dmgH)$_2$(L)

An alternative method for generating CoR(dmgH)$_2$(py) involves the reaction of CoBr(dmgH)$_2$(py) with Grignard reagents:

$$CoBr(dmgH)_2(py) + 2 RMgBr \longrightarrow CoBr(dmgMgBr)_2(py) + 2 RH$$

$$CoBr(dmgMgBr)_2(py) + RMgBr \longrightarrow CoR(dmgMgBr)_2(py) + MgBr_2$$

$$CoR(dmgMgBr)_2(py) + 2 HCl \longrightarrow CoR(dmgH)_2(py) + 2 MgBrCl$$

This method is less attractive because 3 mol of the Grignard reagent are required, the first two moles being consumed in the deprotonation of the acidic OH groups in the dmgH ligands.

In the second step of this experiment, you will use bromine to remove the alkyl group to give $CoBr(dmgH)_2(py)$:

$$CoR(dmgH)_2(py) + Br_2 \longrightarrow RBr + CoBr(dmgH)_2(py)$$

EXPERIMENTAL PROCEDURE

Safety note: Pyridine, bromine, and bromoalkanes are toxic and should only be handled in a hood.

Isopropylbis(dimethylglyoximato)(pyridine)cobalt(III), Co[CH(CH$_3$)$_2$](dmgH)$_2$(py)

The reaction apparatus consists of a 200-mL three-neck round-bottom flask and a pressure-equalizing addition funnel (Fig. 21-2). Fit one neck with a N$_2$ gas inlet

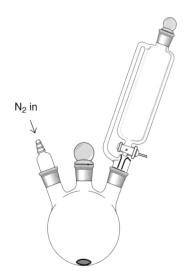

N$_2$ in

Figure 21-2
Apparatus for the synthesis of Co[CH(CH$_3$)$_2$](dmgH)$_2$(py).

but do not place the stopper in the central neck just yet. Lubricate all glass-to-glass joints with silicone grease. Charge the reaction flask with the following: a magnetic stir bar, 75 mL of methanol, 4 g (17 mmol) of $CoCl_2 \cdot 6 H_2O$, and 4 g (34 mmol) of dimethylglyoxime. Charge the pressure-equalizing addition funnel with a solution of 1.25 g (31 mmol) of NaOH in 8 mL of water. While the methanol solution is stirring, flush the entire system with N_2 for several minutes. After reducing the N_2 flow, add the NaOH solution dropwise followed by 1.8 mL (1.75 g, 22 mmol) of pyridine. Stir the mixture for 20 min and then cool it in ice. While the reaction solution is cooling, prepare a solution of 1.5 g (33.5 mmol) of NaOH and 1.5 g (40 mmol) of $NaBH_4$ in 7 mL of H_2O. Add this $NaBH_4$ solution to the cold, stirred reaction mixture. The reaction mixture should become very dark as $Co(dmgH)_2(py)^-$ forms. Finally, add 2.25 mL (2.25 g, 18.3 mmol) of 2-bromopropane to the cold reaction solution, and then place a greased stopper in the central neck and turn off the N_2 flow.* After the mixture has stirred a further 40 min at 0 °C, pour the reaction mixture into a beaker containing 75 mL of water. Collect the orange product by filtration (see Figure 13-1) and dry it under vacuum. Recrystallize your product by suspending the solid in 25 mL of hot (50 °C) methanol in a 100-mL Erlenmeyer flask on a stirring hot plate. Add 5-mL portions of methanol until all of the solid dissolves. At this stage, slowly add 20 mL of water and then allow the solution to cool to room temperature. Collect the product by filtration. Record the yield. Measure the IR spectrum (see Experiment 19 for the preparation of a Nujol mull) and ^1H NMR spectrum of the product.

Bromobis(dimethylglyoximato)(pyridine)cobalt(III), CoBr(dmgH)$_2$(py)

In a three-neck flask equipped with a stir bar, dropping funnel, and a N_2 inlet, dissolve 0.75 g (1.8 mmol) of $Co[CH(CH_3)_2](dmgH)_2(py)$ in 15 mL of acetic acid. Flush the flask with N_2. Add dropwise a solution of 1.2 g (7.5 mmol) of Br_2 in 10 mL of acetic acid. Seal the flask with a stopper, and stir the mixture for a further 1.5 to 2 h. Filter off the green-brown solid product and dry it under vacuum. Determine the yield and record the IR spectrum of the product.

REPORT

Include the following:

1. Yields of $Co[CH(CH_3)_2](dmgH)_2(py)$ and $CoBr(dmgH)_2(py)$.
2. The IR spectra of $Co[CH(CH_3)_2](dmgH)_2(py)$ and $CoBr(dmgH)_2(py)$.
3. The ^1H NMR spectrum of $Co[CH(CH_3)_2](dmgH)_2(py)$.

PROBLEMS

1. What is the valence electron count of the metal center in $Co[CH(CH_3)_2](dmgH)_2(py)$?

---

*The reagents CH_3I, C_2H_5I, or $C_6H_5CH_2Cl$ can be used in place of 2-bromopropane to make the corresponding methyl-, ethyl-, and benzylcobaloxime derivatives.

2. The square planar complex Ni(dmgH)$_2$ reacts with two equivalents of BF$_3$ to give 2 equiv of HF and a complex whose ultraviolet–visible spectrum is similar to the starting complex. Suggest a structure for this product.

3. How do porphyrin and corrin ligands compare with respect to the ring size, formal charge, and degree of unsaturation?

4. It is speculated that cobalt is uniquely suited for its biological role in cobalamins due to the availability of three adjacent oxidation states, Co^{III}, Co^{II}, and Co^{I}. How does this situation compare with the oxidation states utilized by iron in nature? Give examples.

5. Many metalloenzymes have imidazole ligands. Which amino acid bears an imidazole substituent?

INDEPENDENT STUDIES

A. Prepare CoCl(dmgH)$_2$(py) and convert it to its phenyl derivative Co(C$_6$H$_5$)(dmgH)$_2$(py). (Schrauzer, G. N. *Inorg. Synth.* **1968**, *11*, 61.)

B. Prepare methylcobalamin from cyanocobalamin (vitamin B$_{12}$). (Dolphin, D.; Halko, D. J.; Silverman, R. B. *Inorg. Synth.* **1980**, *20*, 134.)

C. Prepare cobaloxime complexes with 4-tert-butylpyridine. (Jameson, D. L.; Grzybowski, J. J.; Hammels, D. E.; Castellano, R. K.; Hoke, M. E.; Freed, K.; Basquill, S.; Mendel, A.; Shoemaker, W. J. *J. Chem. Educ.* **1998**, *75*, 447.)

REFERENCES

Model Complexes of Cobalamins
Bresciani-Pahor, N.; Farcolin, M.; Marzilli, L. G.; Randaccio, L.; Summers, M. F.; Toscano, P. J. *Coord. Chem. Rev.* **1985**, *63*, 1. Influence of axial ligands on the structure and reactivity of cobaloximes.
Franck, B.; Nonn, A. *Angew. Chem., Int. Ed. Engl.* **1995**, *34*, 1795. Novel porphyrinoids via biomimetic synthesis.
Schrauzer, G. N. *Angew. Chem., Int. Ed. Engl.* **1976**, *15*, 417. Review of the cobaloximes by their inventor.
Waddington, M. D.; Finke, R. G. *J. Am. Chem. Soc.* **1993**, *115*, 4629. Study of the thermolysis of an alkylcobalamin with many useful references.

Chemistry and Biochemistry of Cobalamin-Dependent Enzymes
Bannerjee, R. *Chem. Biol.* **1997**, *4*, 175. Review of cobalamin biochemistry.
Brown, K. L.; Evans, D. R.; Zubkowski, J. D.; Valente, E. J. *Inorg. Chem.* **1996**, *35*, 415. The NMR spectra of cobalt corrinoids.
Dolphin, D., Ed. *B$_{12}$, Vol. 1 Chemistry*, Wiley: New York, 1982.
Dolphin, D., Ed. *B$_{12}$, Vol. 2 Biochemistry and Medicine*, Wiley: New York, 1982.
Drennan, C. L.; Huang, S.; Drummond, J. T.; Matthews, R. G.; Ludwig, M. L. *Science* **1994**, *266*, 1669. Structure of methionine synthetase.
Eschenmoser, A. *Angew. Chem., Int. Ed. Engl.* **1988**, *27*, 6. Synthesis and biosynthesis of cobalamins and related nickel enzymes.
Kratky, C.; Farber, G.; Gruber, K.; Wilson, K.; Dauter, Z.; Nolting, H. F.; Konrat, R.; Krauler, B. *J. Am. Chem. Soc.* **1995**, 117, 4654. Structure of aquocobalamin in solution and in the solid state.
Ludwig, M. L.; Matthews, R. G. *Ann. Rev. Biochem.* **1997**, *66*, 269. Review of the structures of B$_{12}$-dependent enzymes.

General References to Bioinorganic Chemistry

Cowan, J. A. *Inorganic Biochemistry*, 2nd ed., Wiley: New York, 1997.

Fraústo da Silva, J. J. R.; Williams, R. J. P. *The Biological Chemistry of the Elements*, Oxford University: Oxford, UK, 1991.

Kaim, W.; Schwederski, B. *Bioinorganic Chemistry: Inorganic Elements in the Chemistry of Life*, Wiley: New York, 1994.

Lehninger, A. L.; Nelson, D. L.; Cox, M. M. *Principles of Biochemistry*, 2nd ed., Worth: New York, 1993.

Lippard, S. J.; Berg, J. M. *Bioinorganic Chemistry*, University Science Books: Mill Valley CA, 1994.

Amino Acid Complexes: Stability Constants of Ni(glycinate)$_n^{(2-n)+}$

Note: This experiment requires one 3- or 4-hour laboratory period.

Biologically important molecules such as amino acids, sugars, and nucleic acids are well known to serve as ligands for transition metals. For example, in many metalloproteins the metal atoms are bound to side chains of amino acid residues (see Experiment 21). There have been extensive investigations of the ability of metal atoms to coordinate to biologically important molecules such as those mentioned above. Chemists have combined their efforts to solve problems that require a knowledge of both inorganic and biochemistry. In this experiment, you will investigate the binding of nickel with glycine, which is one of the 24 commonly occurring amino acids. This study is interesting because several different nickel–glycinate complexes can be present in solution, depending on the concentrations of the reagents and the pH of the solution. By using only a pH meter, however, you will be able to determine the stoichiometries of the species present in solution and the equilibrium constants governing their formation.

First, it is important to discuss what is known about the coordination chemistry of nickel. Although coordination compounds have a variety of geometries, in general the first-row $+2$ transition metal ions react with water to form octahedral aqua complexes of the formula $M(OH_2)_6^{2+}$. The addition of another ligand, L^-, to such solutions results in the competition of L^- for coordination sites occupied by H_2O. If M^{2+} is a labile metal ion, such as Ni^{2+}, Co^{2+}, Cu^{2+}, and Zn^{2+}, the equilibrium

$$M(OH_2)_6^{2+} \; + \; L^- \; \rightleftharpoons \; M(OH_2)_5(L)^+ \; + \; H_2O$$

is established very rapidly. The concentrations of each of the species in solution depend on the ligand properties of L^- and H_2O as well as on their concentrations. Because the concentration of H_2O in dilute solutions is virtually the same regardless of the position of the equilibrium, the factor $[H_2O]$ in the equilibrium expression

$$K \; = \; \frac{[M(OH_2)_5(L)^+][H_2O]}{[M(OH_2)_6^{2+}][L^-]}$$

is effectively a constant and is included in the equilibrium constant. It is frequently written as

$$K_1 = \frac{[M(OH_2)_5(L)^+]}{[M(OH_2)_6{}^{2+}][L^-]}$$

Realizing that H_2O always occupies coordination sites in the complex not
filled by L^-, we can save time by omitting the H_2O ligands from the complex
formulas; for example, $M(OH_2)_5(L)^+$ is written ML^+. Thus, we can write the
following "shorthand" versions of both the equilibrium reaction and the associated equilibrium constant expression:

$$M^{2+} + L^- \rightleftharpoons ML^+ \qquad (1)$$

$$K_1 = \frac{[ML^+]}{[M^{2+}][L^-]} \qquad (2)$$

The constant, K_1, is called a *stability constant* or sometimes a formation
constant.

Although true equilibrium constants are functions of the activities, a, of the
reactants and products,

$$K_\gamma = \frac{a_{ML^+}}{a_{M^{2+}}a_{L^-}} = \frac{[ML^+]}{[M^{2+}][L^-]} \cdot \frac{\gamma_{ML^+}}{\gamma_{M^{2+}}\gamma_{L^-}} \qquad (3)$$

the evaluation of the activity coefficients, γ, is usually difficult and seldom done.
Activity coefficients usually depend on the ionic strength of the solution, but at
infinite dilution γ is equal to 1 and concentrations and activities become equal.
Although it would be desirable to study equilibria in very dilute solutions where
the γ values are known to be 1, in practice this is not always possible. Hence, the γ
values are usually not known, but because equilibrium constants frequently
depend on the ionic strength of the solution, it is at least desirable to maintain a
known and constant ionic strength in the solutions under study. Particularly in
reactions involving anionic ligands L^-, the value of the equilibrium constant will
change with a variation in the ionic strength of the solution. (The ionic strength is
related to the sum of the squares of the electrical charges in solution per unit
volume.)

In most equilibrium studies of metal complex formation, the equilibrium
constant is evaluated from measurements on solutions containing various
concentrations of M^{2+} and L^-. Such concentration changes will produce changes
in the ionic strength of the solutions, in the γ values, and therefore in K_γ. To keep
the ionic strength constant, a large excess of an unreactive ionic salt is added to
the solutions. Then any changes in ionic strength owing to changes in the position
of equilibrium in Eq. 1 will be negligible compared with the high concentration
of the added salt. The salt is added only to maintain the ionic strength of the
medium and should not interact directly with M^{2+} or L^-. Salts such as KNO_3
and $NaClO_4$ have been used extensively because of the low affinity of the $NO_3{}^-$

and ClO_4^- ions for M^{2+}. In this experiment, potassium nitrate, KNO_3, will be used.

Although a large excess of KNO_3 ensures that the ionic strength will be the same for all measurements, the γ values for M^{2+}, L^-, and ML^+ will still be unknown. Because they will not be evaluated, the ratio $\gamma_{ML^+}/\gamma_{M^{2+}}\gamma_{L^-}$ in Eq. 3 will be an unknown constant. The value of K will therefore be calculated using only the concentrations of ML^+, M^{2+}, and L^-, as in Eq. 2. Such a K is called a concentration constant and is valid only at the ionic strength used in the determination. The vast majority of stability constants that are reported in the chemical literature are actually concentration constants.

The existence of six coordination sites in many metal complexes means that L^- and H_2O compete for all six sites, and the following equilibria are possible:

$$M^{2+} + L^- \rightleftharpoons ML^+ \qquad K_1 = \frac{[ML^+]}{[M^{2+}][L^-]}$$

$$ML^+ + L^- \rightleftharpoons ML_2 \qquad K_2 = \frac{[ML_2]}{[ML^+][L^-]}$$

$$ML_2 + L^- \rightleftharpoons ML_3^- \qquad K_3 = \frac{[ML_3^-]}{[ML_2][L^-]}$$

$$ML_3^- + L^- \rightleftharpoons ML_4^{2-} \qquad K_4 = \frac{[ML_4^{2-}]}{[ML_3^-][L^-]}$$

$$ML_4^{2-} + L^- \rightleftharpoons ML_5^{3-} \qquad K_5 = \frac{[ML_5^{3-}]}{[ML_4^{2-}][L^-]}$$

$$ML_5^{3-} + L^- \rightleftharpoons ML_6^{4-} \qquad K_6 = \frac{[ML_6^{4-}]}{[ML_5^{3-}][L^-]}$$

Complexes such as ML_2 and ML_3^-, which represent $ML_2(OH_2)_4$ and $ML_3(OH_2)_3^-$, may exist in cis and trans forms. These equilibrium expressions do not distinguish these isomeric forms. Thus, $[ML_2]$ refers to the total concentration of both isomers of ML_2, and so on. The equilibrium constants $K_1, K_2, \ldots, K_6$ in the series of equilibria are called stepwise stability constants. The equilibrium constants may, for certain purposes (see the next section), also be expressed as overall stability constants, β, which are simply products of the stepwise stability constants.

$$\beta_1 = K_1$$

$$\beta_2 = K_1 K_2$$

$$\beta_3 = K_1 K_2 K_3, \quad \text{and so on}$$

For reactions that involve six separate equilibria (as for $Ni^{2+} + NH_3$), the experimental determination of all six K values is a very difficult, but feasible, task. For this and other reasons, the majority of stability constant investigations have involved reactions of metal ions with chelating ligands. The chelating ligand to be used in this experiment is glycinate $NH_2CH_2CO_2{}^-$, which is the conjugate base of the naturally occurring amino acid, glycine. The glycinate anion coordinates to metal ions through both its nitrogen and oxygen atoms.

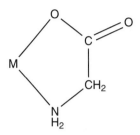

The coordination number of 6 limits the coordination of the metal ion to a maximum of three bidentate glycinate ligands. This experiment is concerned with determining the magnitude of the interaction between $NH_2CH_2CO_2{}^-$ and Ni^{2+}. The equilibrium constants to be determined are shown as follows (the glycinate anion, $NH_2CH_2CO_2{}^-$, is designated A^-):

$$Ni^{2+} + A^- \rightleftharpoons NiA^+ \qquad K_1 = \frac{[NiA^+]}{[Ni^{2+}][A^-]} \qquad (4)$$

$$NiA^+ + A^- \rightleftharpoons NiA_2 \qquad K_2 = \frac{[NiA_2]}{[NiA^+][A^-]} \qquad (5)$$

$$NiA_2 + A^- \rightleftharpoons NiA_3{}^- \qquad K_3 = \frac{[NiA_3{}^-]}{[NiA_2][A^-]} \qquad (6)$$

On the basis of electrostatics, one would expect the affinity of A^- for the complex to decrease as the charge on the complex becomes less positive. It is therefore to be expected that the K values will decrease in the order: $K_1 > K_2 > K_3$.

The experimental determination of the K values for the above equilibria may be carried out by several techniques, but one of the most common is that of measuring with a pH meter the H^+ concentration in a solution containing varying amounts of Ni^{2+} and HA. The H^+ concentration is produced by the ionization of glycine (note that glycine exists in a zwitterionic form):

$$\underset{HA}{{}^+NH_3CH_2CO_2{}^-} \overset{K_a}{\rightleftharpoons} H^+ + \underset{A^-}{NH_2CH_2CO_2{}^-}$$

This equilibrium normally lies far to the left, but addition of Ni^{2+} to the solution results in the release of H^+ depending on the affinity of Ni^{2+} for the chelating $NH_2CH_2CO_2{}^-$:

$$Ni^{2+} + HA \overset{K_e}{\rightleftharpoons} H^+ + NiA^+$$

A knowledge of the concentrations of Ni^{2+} and HA initially added as well as the H^+ concentration (from pH measurement) at equilibrium allows the calculation of the equilibrium constant, K_e, for this reaction. Because $K_e/K_a = K_1$, the value of K_1 may be evaluated. This is the general principle for the determination, but the formation of three Ni–glycine complexes in the present experiment requires that K_1, K_2, and K_3 all be considered simultaneously; their evaluation is significantly more complicated than just mentioned. This evaluation will be discussed in the next section.

Stability constants have been determined for reactions of a large variety of M^{2+} and L^- groups. Numerous correlations of K values with the properties of M^{2+} and L^- have been made. The references at the end of the experiment should be consulted for discussions of this area of research.

pH Titrations of Metal Complexes

As noted above, the acidic character of glycine (HA) allows stability constants for Ni^{2+} complexation of the glycinate ion, $NH_2CH_2CO_2^-$ (A^-), to be evaluated. First, however, the acid dissociation constant, K_a, of glycine, must be determined.

$$HA \rightleftharpoons H^+ + A^-$$

$$K_a = \frac{[H^+][A^-]}{[HA]} \tag{7}$$

The value of K_a can be determined by measuring $[H^+]$ in a solution of known HA concentration. The proton concentration $[H^+]$ is commonly measured with a pH meter and a standard glass electrode. Because the pH meter determines the electromotive force (emf) or voltage of the glass electrode relative to a reference electrode, a pH measurement determines the H^+ activity, a_{H^+}, and not its concentration, $[H^+]$, in solution. Because we want to calculate concentration stability constants, it will be necessary to convert a_{H^+} to $[H^+]$ by using the known activity coefficient of H^+ in a solution having a given ionic strength, μ. The mean activity coefficient, $\gamma_\pm$, of the ions in, for example, HNO_3, may be calculated from the Davies equation, which is an empirical extension of the Debye–Hückel limiting activity equation:

$$-\log \gamma_\pm = \frac{0.50\, Z_1 Z_2 \mu^{1/2}}{1 + \mu^{1/2}} - 0.10\, \mu \tag{8}$$

In this expression, Z_1 and Z_2 are the charges $+1$ and -1 on H^+ and NO_3^-, respectively. The ionic strength of the solution, μ, is given by the usual definition,

$$\mu = \tfrac{1}{2} \sum_i M_i Z_i^2$$

where M_i is the molar concentration of ion i and Z_i is its charge. The ionic strength of the solutions used in this experiment will be determined virtually entirely by the KNO_3 concentration.

By definition

$$a_{H^+} = \gamma_{\pm}[H^+]$$

Rearranging and substituting $pH = -\log a_{H^+}$ gives

$$\log[H^+] = -pH - \log(\gamma_{\pm}) \qquad (9)$$

Hydroxide ion concentrations may also be calculated from pH measurements by making use of the autoionization constant, $K_w = [H^+][OH^-]$, of water ($K_w = 1.615 \times 10^{-14}$ at 25.0 °C and 0.10 M ionic strength):

$$\log[OH^-] = pH - pK_w + \log(\gamma_{\pm}) \qquad (10)$$

Hence, it is possible to evaluate $[H^+]$ and $[OH^-]$ for use in calculating concentration equilibrium constants from pH measurements.

Returning to the problem of evaluating K_a for glycine, suppose a solution of glycine in 0.10 M KNO_3 is titrated with NaOH. After each addition of NaOH, the pH of the solution is measured. In such solutions, several conditions must hold.

1. The concentration of positive charge must equal that of negative charge, that is,

$$[H^+] + [Na^+] = [OH^-] + [A^-]$$

2. The total glycine concentration, A_{tot}, of the prepared solution must be

$$A_{tot} = [HA] + [A]$$

3. The expression for the acid dissociation constant, Eq. 7, pertains.

Combination of these three expressions yields the following equation for the pK_a, $-\log K_a$, for glycine:

$$pK_a = -\log[H^+] + \log\left(\frac{A_{tot} - ([Na^+] + [H^+] - [OH^-])}{[Na^+] + [H^+] - [OH^-]}\right) \qquad (11)$$

Because all quantities on the right-hand side of this expression either are known or may be calculated from Eqs. 9 and 10, the pK_a of glycine may be evaluated. The value found experimentally is 9.60 (so that $K_a = 2.5 \times 10^{-10}$).

Now the method for determining the stepwise stability constants K_1, K_2, and K_3, for the reaction of Ni^{2+} with A^- may be introduced. Experimentally, a solution containing 1 mmol of Ni^{2+} (as $NiCl_2 \cdot 6 H_2O$) and 1 mmol of H^+ (as HNO_3) will be titrated with a solution of sodium glycinate, $NH_2CH_2CO_2^- Na^+$, prepared by neutralizing glycine with NaOH. This titration will give a solution containing an equilibrium mixture of H^+, OH^-, Na^+, HA, A^-, Ni^{2+}, NiA^+, NiA_2, and NiA_3^-. From pH measurements and a knowledge of the quantities of Ni^{2+}, H^+, HA, and $NaOH$ originally added, it is possible to calculate the stability constants.

The method to be used was largely developed by J. Bjerrum. To facilitate the determination of the K values, a function, $\bar{n}$, is defined as the average number of ligand molecules bound per metal ion. For the present system

$$\bar{n} = \frac{\text{moles of bound } A^-}{\text{total moles of } Ni^{2+}} = \frac{[NiA^+] + 2[NiA_2] + 3[NiA_3^-]}{[Ni^{2+}] + [NiA^+] + [NiA_2] + [NiA_3^-]}$$

Substituting from Eqs. 4–6 gives the expression

$$\bar{n} = \frac{K_1[A^-] + 2K_1K_2[A^-]^2 + 3K_1K_2K_3[A^-]^3}{1 + K_1[A^-] + K_1K_2[A^-]^2 + K_1K_2K_3[A^-]^3} \tag{12}$$

From an experimental standpoint, $\bar{n}$ may be expressed in terms of the total glycine concentration (A_{tot}), the concentration of HA and A^-, and the total Ni^{2+} concentration (M_{tot}):

$$\bar{n} = \frac{A_{tot} - [HA] - [A^-]}{M_{tot}} \tag{13}$$

To determine $[HA]$ and $[A^-]$ in Eq. 13, an expression for the H^+ bound to the glycinate ion is introduced:

$$\text{Bound } H^+ = [HA] = \left(\begin{array}{c}\text{added } H^+ \\ \text{from } HNO_3\end{array}\right) + \left(\begin{array}{c}H^+ \text{ from} \\ \text{dissociation} \\ \text{of } H_2O\end{array}\right) - \text{free } H^+$$

$$[HA] = C_H + [OH^-] - [H^+] \tag{14}$$

where C_H is the concentration of $[H^+]$ in the Ni^{2+} solution due to the added HNO_3. Substitution of Eq. 7 gives

$$[A^-] = \frac{K_a}{[H^+]}(C_H + [OH^-] - [H^+]) \tag{15}$$

Substitution of Eqs. 14 and 15 into Eq. 13 yields

$$\bar{n} = \frac{A_{tot} - (1 + K_a/[H^+])(C_H + [OH^-] - [H^+])}{M_{tot}} \tag{16}$$

Thus, $[A^-]$ and $\bar{n}$ may be calculated from experimentally known quantities.

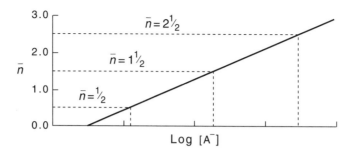

Figure 22-1
Typical plot of the average number of ligands per metal ion, $\bar{n}$,
versus the concentration of the free ligand, A^-.

In Figure 22-1 is shown a typical plot of the average number of ligands per metal ion, $\bar{n}$, versus the concentration of the free ligand, A^-. From such plots, it is possible to estimate the values of K_1, K_2, and K_3. From Eq. 4, for example, it is noted that when $[NiA^+] = [Ni^{2+}]$ then $K_1 = 1/[A^-]$. Because the condition $[NiA^+] = [Ni^{2+}]$ means that the average number of ligands per metal ion is $\frac{1}{2}$, that is, $\bar{n} = \frac{1}{2}$, the value of $[A^-]$ at $\bar{n} = \frac{1}{2}$ from a plot as in Figure 22-1 will allow K_1 to be estimated. Likewise at $\bar{n} = 1\frac{1}{2}$ and $2\frac{1}{2}$, the values of $[A^-]$ will allow the estimation of K_2 and K_3. In general, K_n can be estimated from the expression: $K_n = [1/[A^-]]_{\bar{n}=n-\frac{1}{2}}$ Note that these K_n values are only estimates because more than two complexes, for example, more than just NiA^+ and Ni^{2+}, are generally present in a solution for which $\bar{n} = n - \frac{1}{2}$.

A more precise graphical method of evaluating K_1, K_2, and K_3 from $\bar{n}$ and $[A^-]$ data has been developed by Rossotti and Rossotti and will be used in this experiment.

First, Eq. 12 is converted to Eq. 17 by introducing expressions for the overall stability constants, β:

$$\bar{n} = \frac{\beta_1[A^-] + 2\beta_2[A^-]^2 + 3\beta_3[A^-]^3}{1 + \beta_1[A^-] + \beta_2[A^-]^2 + \beta_3[A^-]^3} \tag{17}$$

where $\beta_1 = K_1$, $\beta_2 = K_1K_2$, and $\beta_3 = K_1K_2K_3$. Rearrangement of this equation gives

$$\frac{\bar{n}}{(1 - \bar{n})[A^-]} = \beta_1 + \frac{(2 - \bar{n})[A^-]}{(1 - \bar{n})}\beta_2 + \frac{(3 - \bar{n})[A^-]^2}{(1 - \bar{n})}\beta_3 \tag{18}$$

Thus, a plot of $\bar{n}/(1 - \bar{n})[A^-]$ versus $(2 - \bar{n})[A^-]/(1 - \bar{n})$ tends to a straight line at low $[A^-]$ of intercept β_1 and slope β_2 (Fig. 22-2).

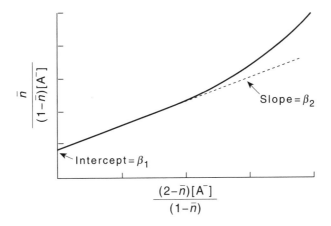

Figure 22-2
Plot to determine the overall stability constant β_1.

By using the value of β_1 obtained as the intercept, Eq. 18 is divided by $(2 - \bar{n})[A^-]/(1 - \bar{n})$ and rearranged to give

$$\frac{\bar{n} - (1 - \bar{n})\beta_1[A^-]}{(2 - \bar{n})[A^-]^2} = \beta_2 + \frac{(3 - \bar{n})}{(2 - \bar{n})}[A^-]\beta_3 \qquad (19)$$

The intercept of a plot of

$$\frac{\bar{n} - (1 - \bar{n})\beta_1[A^-]}{(2 - \bar{n})[A^-]^2} \quad \text{versus} \quad \frac{3 - \bar{n}}{2 - \bar{n}}[A^-]$$

gives β_2 and the slope is β_3 (Fig. 22-3).

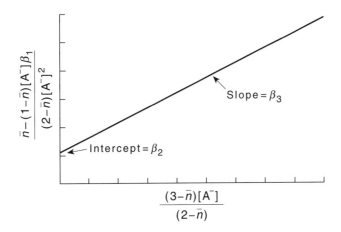

Figure 22-3
Plot to determine the overall stability constants β_2 and β_3.

From these relatively precise values of β_1, β_2, and β_3, the stability constants K_1, K_2, and K_3 may be calculated. Because of the method of data treatment, the precisions of the constants decrease in the order $K_1 > K_2 > K_3$.

Calculations of the functions required for the evaluation of the stability constants are lengthy and involved. We strongly recommend that you carry out the calculations on a computer. See the references at the end of this experiment.

EXPERIMENTAL PROCEDURE

A pH meter is a sensitive instrument that must be operated properly and with care. The electrodes are particularly easy to break or crack. When not in use, they should be stored in a buffer solution of approximately pH 4. Just before use, they should be rinsed with distilled water (use a wash bottle) and blotted dry with a piece of filter paper. They may then be introduced into the solution under study.

The pH meter must first be standardized. Insert the electrodes into a pH 7 buffer, and if necessary adjust the meter until it reads the same as the pH of the buffer. Then, using buffers of pH 4 and 10, check the meter reading. If the reading differs by more than 0.1 pH unit from that of the buffer, recheck the meter reading of the pH 7 buffer. If the adjustment for pH is correct at pH 7, yet the reading is incorrect at pH 4 and 10, the meter reading will have to be corrected for this inaccuracy. This correction may be done on some pH meters by adjusting the slope adjustment or the temperature compensator. Otherwise, it should be accomplished by determining a meter correction constant, C_{meter}, from the linear expression:

$$ pH = 7.00 + (pH_{meter} - 7.00)C_{meter} \qquad (20) $$

where pH is the correct pH, and pH_{meter} is the meter reading. Using the pH 4 or 10 buffer, determine C_{meter}. To obtain correct pH readings in all subsequent measurements, pH_{meter} readings will have to be corrected with Eq. 20.

Prepare the following solutions, using distilled water:

> 20 mL 0.4 M glycine, HA
> 500 mL 0.2 M KNO$_3$
> 100 mL 0.10 M HNO$_3$
> 100 mL 0.50 M NaOH

Standardize the NaOH solution by titrating a solution of the primary standard, potassium acid phthalate, using phenolphthalein indicator. Use the standardized NaOH solution to standardize the HNO$_3$ solution. All solutions (particularly NaOH) should be stored in tightly stoppered bottles to prevent atmospheric CO$_2$ from dissolving to give H$_2$CO$_3$, H$^+$, HCO$_3^-$, and CO$_3^{2-}$.

Next, the data necessary for calculating K_1, K_2, and K_3 will be collected. Prepare 20 mL of 0.4 M sodium glycinate solution by exactly neutralizing weighed solid glycine with a calculated amount of standardized 0.5 M NaOH and diluting with water to a total volume of 20 mL. Rinse and fill the cleaned 10-mL buret with this solution.

In a 400-mL beaker, prepare a solution of 100 mL of 0.2 M KNO_3, 0.24 g (1.0 mmol) of $NiCl_2 \cdot 6 H_2O$, 10 mL of 0.10 M HNO_3, and 90 mL of distilled H_2O. Using the sodium glycinate solution, titrate this solution with stirring. After each 0.2-mL aliquot addition, record the amount of glycinate solution added and the pH of the solution. Continue adding 0.2-mL aliquots until you have added the full 10 mL. Because the volume (10 mL at the end of the titration) of sodium glycinate solution added is small compared to the volume (200 mL) of the Ni^{2+} solution, correcting the concentrations for dilution is not necessary. From the titration data, calculate $[A^-]$ from Eq. 15 and $\bar{n}$ from Eq. 16 at each of the 50 points. Note that $[H^+]$ and $[OH^-]$ may be negligible in some cases. In these calculations, be sure to correct the meter reading (Eq. 20), if necessary, and to convert pH to $[H^+]$ using Eqs. 9 and 10. Make a plot of $\bar{n}$ versus $\log [A^-]$ as in Figure 22-1. From the values of $[A^-]$ at $\bar{n} = \frac{1}{2}$, $1\frac{1}{2}$, and $2\frac{1}{2}$, calculate approximate values of K_1, K_2, and K_3.

Now calculate

$$\frac{\bar{n}}{(1 - \bar{n})[A^-]} \quad \text{and} \quad \frac{(2 - \bar{n})[A^-]}{(1 - \bar{n})}$$

for those points with values of $\bar{n}$ between 0.2 and 0.8. Plot these quantities as in Figure 22-2 to obtain an accurate value of β_1 from the intercept. Then evaluate

$$\frac{\bar{n} - (1 - \bar{n})\beta_1[A^-]}{(2 - \bar{n})[A^-]^2} \quad \text{and} \quad \frac{(3 - \bar{n})}{(2 - \bar{n})}[A^-]$$

for the points whose $\bar{n}$ values range from 1.1 to 1.7. From the plot of these quantities as in Figure 22-3, determine β_2 (intercept) and β_3 (slope). Finally, calculate K_1, K_2, and K_3 from β_1, β_2, and β_3 and compare these accurate values with the approximate ones that you evaluated earlier by using the half-$\bar{n}$ method.

REPORT

Include the following:
1. Plot of $\bar{n}$ versus $\log [A^-]$. Approximate K_1, K_2, and K_3 values from half-$\bar{n}$ method.
2. Plot used to determine β_1.
3. Plot used to determine β_2 and β_3.
4. Accurate values of K_1, K_2, and K_3 in 0.10 M KNO_3.
5. Comment on the relative precisions of the K_1, K_2, and K_3 values.

PROBLEMS

1. Draw all geometrical and optical isomers of the three complexes $Ni(glycinate)^+$, $Ni(glycinate)_2$, and $Ni(glycinate)_3^-$.
2. Derive Eq. 11.

3. A solution prepared by mixing 100 mL of 0.2 M glycine and 100 mL of 0.2 M Cu^{2+} is adjusted to a pH of 7 with NaOH. Calculate approximate percentages of the original Cu^{2+} that exists as Cu^{2+}, CuA^+, and CuA_2 in the solution. Use a pK_a of 9.60 for glycine, and the following stability constants:

$$Cu^{2+} + A^- \rightleftharpoons CuA^+ \qquad \log K_1 = 8.38$$

$$CuA^+ + A^- \rightleftharpoons CuA_2 \qquad \log K_2 = 6.87$$

4. You wish to determine the formation constants (K_1, K_2, and K_3) for the coordination of ethylenediamine, $NH_2CH_2CH_2NH_2$ (en), to Ni^{2+} to form $Ni(en)^{2+}$, $Ni(en)_2^{2+}$, and $Ni(en)_3^{2+}$. (a) How would you determine the first and second pK_a values of en? (b) Define these two pK_a values. Which of the two will be larger, and why? (c) Briefly explain what titrations you would carry out in order to determine the stability constants K_1, K_2, and K_3. Would the procedure differ from that used in this experiment? (d) Which value would you expect to be larger, K_1 or K_2?

5. Consider the coordination of Cu^{2+} by the tridentate iminodiacetate ligand, $HN(CH_2CO_2^-)_2$, abbreviated A^{2-}. When the monoprotonated form, $H_2N(CH_2CO_2^-)_2$, abbreviated HA^-, is added to an aqueous solution of Cu^{2+}, the following equilibrium is rapidly established:

$$Cu^{2+} + HA \rightleftharpoons CuA + H^+$$

If a solution prepared by combining 100 mL of 0.02 M Cu^{2+} and 100 mL of 0.02 M HA^- is adjusted to pH 1.60 with NaOH, what are the concentrations of Cu^{2+}, HA^-, and CuA in the solution? (The K_a for HA^- is 4.70×10^{-10}, and K_1 for $Cu^{2+} + A^- \rightleftharpoons CuA^+$ is 4.26×10^{10}. Draw the probable structure of the CuA complex.

6. The pH method of measuring equilibrium constants is a bulk technique in that it does not discriminate between geometrical isomers. Suggest techniques that would permit separate detection of isomers.

INDEPENDENT STUDIES

A. Determine stability constants for the coordination of Ni^{2+} by another amino acid, such as alanine or valine.

B. Determine stability constants for the coordination of glycine at a higher ionic strength, such as 1.0 M KNO_3. Compare the results with those at 0.1 M KNO_3.

C. Determine stability constants for the coordination of Cu^{2+} (or other metal ions) by glycine.

D. Using the method of continuous variations (Job's method), establish the compositions of the $Ni(glycinate)_n^{(2-n)+}$ complexes present in solutions of Ni^{2+} and glycinate.

E. Determine stability constants for the stepwise coordination of Cu^{2+} by NH_3. (Guenther, W. B. *J. Chem. Educ.* **1967**, *44*, 46.)

F. Determine stability constants for the stepwise coordination of Cu^{2+} by ethylenediamine. (Goldberg, D. E. *J. Chem. Educ.* **1962**, *39*, 328.)

G. Determine pK_a values of trifunctional amino acid, cysteine. (Clement, G. E.; Hartz, T. P. *J. Chem. Educ.* **1971**, *48*, 395.)

REFERENCES

$Ni(glycinate)_n^{(2-n)+}$ and other Amino Acid Complexes of Transition Metals

Braibanti, A.; Ostacoli, G.; Paoletti, P.; Pettit, L. V.; Sammartano, S. *Pure Appl. Chem.* **1987**, *59*, 1721. Stability constants for nickel glycinate.

Li, N. C.; White, J. M.; Yoest, R. L. *J. Am. Chem. Soc.* **1956**, *78*, 5218. Use of the Bjerrum method to study nickel glycinate.

Nancollas, G. H.; Tomson, M. B. *Pure Appl. Chem.* **1982**, *54*, 2676. Stability constants for nickel glycinate.

Stability Constant Determination Techniques and Data Treatment

Barbosa, J.; Barrón, D.; Beltrán, J. L.; Sanz-Nebot, V. *Anal. Chim. Acta* **1995**, *317*, 75. A computer program for the refinement of equilibrium constants from titration curves.

Lomozik, L.; Jaskólski, M.; Gasowska, A. *J. Chem. Educ.* **1995**, *72*, 27. Comparison of three different computer programs for the calculation of stability constants.

Martell, A. E.; Motekaitis, R. J. *Determination and Use of Stability Constants*, 2nd ed., VCH: New York, 1992. Useful guide with a diskette containing software programs for analyzing titration data.

Meloun, M.; Havel, J.; Hogfeldt, E. *Computation of Solution Equilibria: A Guide to Methods in Potentiometry, Extraction, and Spectrophotometry*, Halsted: New York, 1988.

Whisenhunt, D. W.; Neu, M. P.; Hou, Z. G.; Xu, J.; Hoffman, D. C.; Raymond, K. N. *Inorg. Chem.* **1996**, *35*, 4128. Stability constant measurements on new multidentate ligands potentially useful for treating patients exposed to plutonium or thorium.

Stability Constants of Complexes

Hinton, J. F.; Amis, E. S. *Chem. Rev.* **1971**, *71*, 627.

Lincoln, S. F. *Coord. Chem. Rev.* **1971**, *6*, 309. Solvent coordination numbers of metal ions in solution.

Martell, A. E.; Hancock, R. D. *Metal Complexes in Aqueous Solution*, Plenum: New York, 1996.

Perrin, D. D., Ed. *Stability Constants of Metal–Ion Complexes, part B, Organic Ligands*, IUPAC Chemical Data Series No. 22, Pergamon: Oxford, UK 1979. A thorough compilation of stability constants of metal ions with inorganic and organic ligands.

Smith, R. M.; Martell, A. E. *Critical Stability Constants*, Plenum Press: New York, 1974–1989. A six-volume critical review of stability constants.

For General References to Bioinorganic Chemistry, see Experiment 21

Bioinorganic Coordination Chemistry: Copper(II) Tetraphenylporphyrinate

Note: This experiment requires 2 hours for Part A and 6 hours for Part B.

Metal ions play vital roles in many biological processes, and at least seven transition metals (iron, zinc, copper, manganese, cobalt, nickel, and molybdenum) are essential to almost all life on earth. These metals are key components of many important proteins. In some cases, the metals coordinate to the nitrogen, sulfur, or oxygen atoms in the side chains of certain amino acids that make up the protein's structure; among these "metal-binding" amino acids are histidine, cysteine, methionine, tyrosine, aspartic acid, and glutamic acid. In other cases, however, the transition metals are bound to special ligands, the most important of which are the porphyrins. Metal-bearing porphyrin complexes are called metalloporphyrins.

Porphyrins have the general structure shown in Figure 23-1; they are compounds with a central 16-membered ring consisting of four pyrrole subunits linked by one-carbon bridges. The porphyrin ring is polyunsaturated and completely conjugated; consequently, porphyrins and their complexes with transition metals are intensely colored. In metalloporphyrins, a metal atom coordinates

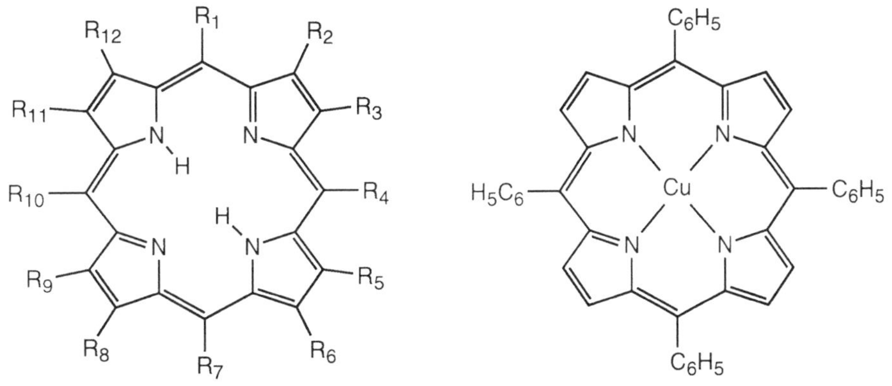

Figure 23-1
General structure of a porphyrin with peripheral groups R_1–R_{12}; structure of the copper(II) complex Cu(TPP).

to the four nitrogen atoms and displaces the two central hydrogen atoms. One important metalloporphyrin complex that is found in almost all animals is hemoglobin, which contains four iron-porphyrin units. In vertebrates, hemoglobin is responsible for the transport of O_2 from the lung to cells throughout the body. Metalloporphyin complexes perform a variety of other important biochemical functions; for example, they serve as electron-transfer relays and as oxidation catalysts.

Closely related to metalloporphyrins are the chlorins, corrins, and corphins, which have similar though not identical structures. Two important examples of such porphyrin-like molecules are chlorophyll and vitamin B_{12} (see Experiment 21). Phthalocyanines, which are nitrogen-rich analogues of porphyrins, are produced and used industrially as pigments and catalysts.

Naturally occurring porphyrins generally have a variety of different organic groups on the periphery (exterior) of the ring. The synthesis of such polysubstituted porphyrin rings is a challenging task for the chemist. For many purposes, however, the characteristic chemical properties associated with metalloporphyrins are exhibited by simpler analogues with small peripheral groups. The most important examples of such synthetic analogues are the complexes of *meso*-tetraphenylporphyrin (abbreviated H_2TPP), where the "meso" designation means that the phenyl groups are located on the four carbon atoms that bridge between the pyrrole rings. All the other peripheral groups in H_2TPP are hydrogen atoms, and the chemical formula of H_2TPP is $C_{44}H_{30}N_4$. In this experiment, you will prepare H_2TPP and convert it to its copper complex $Cu(TPP)$, whose structure is shown in Figure 23-1.

This experiment begins with the preparation of H_2TPP by the condensation of four molecules each of benzaldehyde and pyrrole. This reaction does not give a high yield but the starting materials are inexpensive and the product is easily isolated. The low yields of this reaction illustrate the difficulty of assembling a large ring in a "one-pot" reaction. In some cases, but not this one, the yields of such cyclization reactions can be improved by adding metal ions to the solution, which assist in the assembly of the macrocycle by binding to the reactants and orienting them in such a fashion to favor formation of the ring.

The stoichiometry of the reaction of benzaldehyde and pyrrole is as follows:

$$4\ C_6H_5CHO\ +\ 4\ C_4H_4NH\ +\ \tfrac{3}{2}\ O_2\ \longrightarrow\ C_{44}H_{30}N_4\ +\ 7\ H_2O$$

The mechanism of the cyclization reaction is known in some detail: It involves formation of a carbocation by addition of a proton to benzaldehyde, followed by electrophilic attack of the carbocation at the α position of pyrrole. Loss of water generates a new carbocation, which attacks a second pyrrole ring. These steps are repeated, and eventually a nonconjugated macrocycle is formed. Oxidation of this macrocycle by O_2 generates the fully conjugated porphyrin ring.

Because the oxidation step is not quantitative, however, one of the principal contaminants is *meso*-tetraphenylchlorin (H_2TPC), a compound that contains two more hydrogen atoms than H_2TPP. The chemical structure of H_2TPC is shown in Figure 23-2; the extra hydrogen atoms are present on the β-carbon atoms of one of the pyrrole rings:

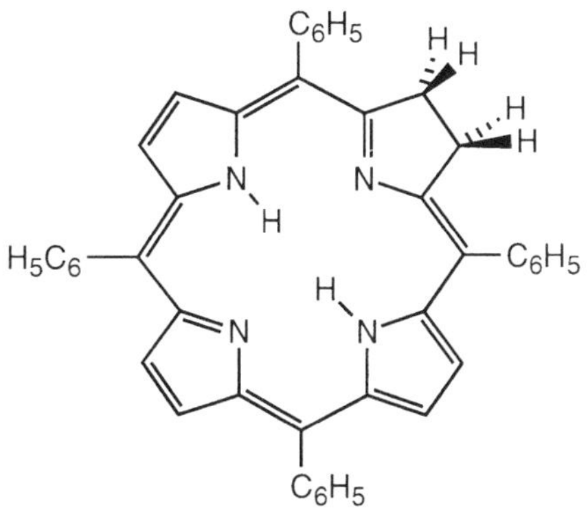

Figure 23-2
Structure of *meso*-tetraphenylchlorin (H$_2$TPC).

Tetraphenylporphyrin and its metal complexes are best separated from tetraphenylchlorin impurities by chromatography.

The conversion of H$_2$TPP to Cu(TPP) is achieved by the addition of an excess of copper(II) acetate. The copper(II) center replaces the two nitrogen-bound protons at the center of the H$_2$TPP ring:

$$H_2TPP \ + \ Cu(O_2CCH_3)_2(H_2O) \ \longrightarrow \ Cu(TPP) \ + \ 2 \ HO_2CCH_3 \ + \ H_2O$$

The insertion of a metal ion into the porphyrin ring is sometimes very slow even though the reaction is very favorable thermodynamically. This somewhat surprising behavior reflects the rigidity and steric constraints of the macrocycle, which make it difficult for the porphyrin nitrogen atoms to approach a metal center that is already surrounded by other ligands (such as water or acetate groups).

Thin-Layer and Column Chromatography
One of the goals of this experiment is to illustrate standard chromatographic techniques that are used in the isolation of pure compounds from complex reaction mixtures. Thin-layer chromatography (TLC) will be used to test the purity of small amounts of the Cu(TPP) sample and to explore the chromatographic conditions necessary to purify it. Once these conditions are established, column chromatography will be utilized to separate larger amounts of Cu(TPP) from copper(II) tetraphenylchlorin and other impurities.

Separations effected by both TLC and column chromatography are based on the tendency of molecules to bind to certain solids called adsorbents. The adsorbents most frequently used are silica gel and alumina. Silica gel, which will be used as the adsorbent in this experiment, is to a first approximation a hydrated form of silicon dioxide, SiO$_2 \cdot x$ H$_2$O. It does, however, contain significant amounts of other inorganic salts whose amounts vary from one silica gel preparation to another. For TLC, the silica gel is often mixed with plaster to help bind the gel to a glass support.

When silica gel is heated in strongly acidic or basic solutions, it acquires acidic or basic properties. Acid-treated silica gel strongly adsorbs (or binds) basic compounds such as amines, whereas base-treated silica gel adsorbs acidic compounds. The adsorption properties of silica gels also depend on their water content. If the gels are strongly heated under vacuum, water is driven off the silica gel leaving sites where other polar molecules strongly adsorb. Less strongly adsorbing silica gel can be prepared by adding back small amounts of H_2O to occupy some of the adsorption sites. By altering the water content, it is therefore possible to control the degree to which silica gel binds various compounds.

In TLC, a thin layer (0.1–2 mm thick) of the adsorbent is spread onto a flat surface. The TLC plates can be purchased commercially and can also be prepared by coating microscope slides. A small amount of the sample to be separated is dissolved in a small volume of a suitable solvent. It is essential that the sample to be tested be completely dissolved. With a capillary, a spot (3–5 mm in diameter) of the solution is placed about 8 mm from one end of the TLC plate. The plate is allowed to dry and then the procedure is repeated to add sample to the same spot. The spot should be kept small for maximum separation of the components. The TLC plate is then placed in a bottle that contains a few milliliters of solvent (Fig. 23-3). The solvent level must be lower than the spot on the

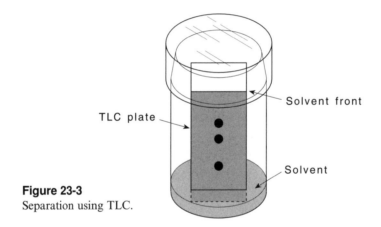

Figure 23-3
Separation using TLC.

plate. The bottle is capped and left undisturbed while the solvent rises up the silica gel by capillary action. If the proper solvent is chosen, the mixture will begin to move up the plate behind the solvent front. Ideally, different compounds present in the mixture will move up the plate at different rates, and thus will be separated.

When the solvent has moved about three-fourths of the way up, the plate should be removed from the bottle. If the compounds in the mixture are colored, it will be obvious if a separation has occurred. If some of the compounds are colorless, their locations on the TLC plate can usually be established by placing the air-dried TLC plate in a bottle containing a few crystals of iodine. The iodine sublimes and adsorbs in the areas where the compounds are located. Thus, dark brown spots on the plate indicate the locations of the components of the original sample. Alternatively, some TLC plates contain a fluorescent indicator:

Upon illumination with an ultraviolet (UV) light, the air-dried plate will glow in all places except the spots where compounds are located.

The separation of mixtures into their components by chromatography mainly depends on differences in the adsorption tendencies and solubilities of the components. Compounds that are weakly adsorbed and that are readily soluble in the solvent will move or elute quickly; in contrast, compounds that are strongly adsorbed and that are poorly soluble in the solvent will elute slowly. Finding the solvent that effects the best separation of the components is not easy, but generally a solvent that dissolves the desired compound moderately well will allow the compound to move up the plate; it can only be hoped that the impurities do not migrate at the same rate as does the compound of interest. If the solvent that was chosen does not separate the components of the mixture, other solvents either more or less polar than the first should be tried until a solvent that gives a separation is found. The polarities of some common chromatographic solvents increase in the following order:

$$\text{increasing polarity} \downarrow \quad \begin{array}{l} \text{alkanes} \\ \text{toluene} \\ \text{benzene} \\ \text{dichloromethane} \\ \text{chloroform} \\ \text{diethyl ether} \\ \text{ethyl acetate} \\ \text{acetone} \\ \text{ethanol} \\ \text{methanol} \\ \text{water} \end{array}$$

It is sometimes convenient to use mixtures of solvents. For example, mixtures of ethyl acetate in dichloromethane often succeed in effecting a useful separation where other solvents do not.

At this point it is probably obvious that the successful choice of adsorbent and solvent is an art that is learned largely by doing chromatographic separations. The references at the end of this experiment do offer, however, many hints on how to use these techniques effectively.

Having established the solvent or solvent mixture that will separate the sample on TLC plates, it is hoped that the same solvent can be used to separate larger quantities of the sample on a silica gel chromatography column. Generally, this is possible. It is necessary, however, to use a much larger silica gel particle size for column chromatography (80–200 mesh) than that used in TLC (finer than 200 mesh).

The chromatography column (Fig. 23-4) consists of a glass tube that typically is about 3 cm in diameter and 30 cm long. First, a small glass wool plug is pushed to the bottom of the column, and a 5-mm layer of sand is added. A slurry of the silica gel in the solvent to be used in the separation is then poured onto the sand. The column is drained until the solvent level is the same as the top of the silica gel, and then a mixture of the sample and silica gel (both suspended in a few milliliters of the solvent) is added to the column. (Alternatively, the sample may

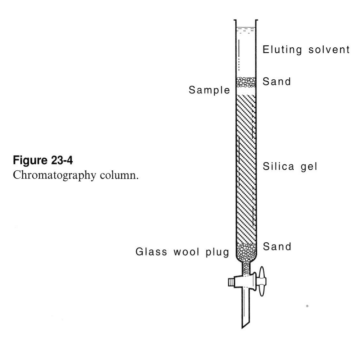

Figure 23-4
Chromatography column.

be dissolved in a small volume of the solvent and added to the column.) A 5-mm layer of sand is then added to the top of the silica gel.

From this point on, the solvent level should never fall below the top of the silica gel, because channels in the column will result, and the solution will pass down the channels without properly percolating through the adsorbent. The solvent is added to the column, and the movement of the compounds down the column begins. This process is called *elution*. The rate at which solvent is passed through the column is called the flow-rate; slow flow-rates give better separations than do fast flow-rates.

If the compounds in the sample are colored, it is easy to determine which fractions should be collected to obtain the desired products. If the compounds are colorless, fractions may have to be collected at regular volume intervals and examined for the presence of the desired compounds spectrophotometrically or by other techniques. Whereas the initial eluting solvent may elute one or more of the compounds, other compounds may require (as noted by TLC) more polar solvents to move them down the column. If a change in solvent is required, it is best to introduce it gradually by using first a mixture of the initial and the subsequent solvent and then finally the pure new solvent. Sometimes an abrupt change of solvent leads to the evolution of large amounts of heat when the new solvent adsorbs to the silica gel. Thermal expansion of the solvent creates channels in the adsorbent that can destroy the efficiency of the column.

The fractions eluted from the column that contain the desired compounds may simply be evaporated to dryness to give the pure product. Evaporation sometimes does not give a crystalline solid, and in such cases recrystallization of the material usually gives purer and better looking product.

Column chromatography has numerous variations. ... alumina (Al_2O_3) are the most common adsor... used. For materials that decompose at roc... separations can be carried out in cooled, jack... pounds have been chromatographed in an atmo... of these variations, however, are basically chron... that will be practiced in this experiment.

EXPERIMENTAL PROCEDURE

Part A

insert Claisen adapter b/t flask & condenser

meso-Tetraphenylporphyrin, $C_{44}H_{30}N_4$ (H$_2$TPP)

Place a Teflon-coated stirbar in a 100-mL one-neck rou... 40 mL of propanoic acid (sometimes called propionic ... ... with a reflux condenser (see Fig. 23-5), and use a small amount ...one grease on the joint. Leave the top of the reflux condenser open to air. Heat the acid to reflux with a rheostat-controlled heating mantle. When the propanoic acid begins to boil vigorously, add a mixture of 1.65 mL (15.75 mmol) of benzaldehyde and 1.0 mL (14.4 mmol) of pyrrole by pouring this solution down the reflux condenser. Measure out these liquids with a syringe or pipet. (*Note:* Freshly distilled pyrrole gives a higher yield of product but is not required.) Rinse the pyrrole and ben-zaldehyde down the condenser with 10 mL of propanoic acid. Continue to reflux the solution for 30 min, and then remove the heat and let the flask cool for a few minutes. Filter the dark brown mixture through a medium-porosity glass frit (see Figure 13-1). Rinse the mixture with a few mL of methanol until the washings are clear and purple crystals are left on the frit. Allow the crystals to dry by pulling air through them for a few minutes. Collect the purple crystals (do not scrape the

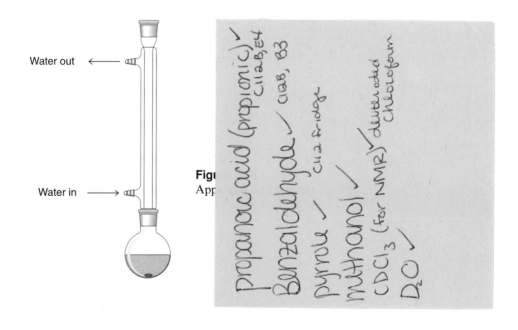

Water out ←

Water in →

Figu...
App...

propanoic acid (propionic) ✓
C11a,B,E4
Benzaldehyde ✓ c11a,B, B3
c11a Fridge
pyrrole ✓ c11a Fridge
methanol ✓
CDCl$_3$ (for NMR) ✓ deuterated chloroform
D$_2$O ✓

frit too vigorously or your sample will become contaminated with powdered glass). Record the yield. Discard the wash solutions by pouring them into the containers provided for them.

Record the proton nuclear magnetic resonance (^{1}H NMR) spectrum of your product. To do this, place approximately 15 mg of your sample in a 5-mm diameter NMR tube and add about 1 mL of CDCl$_3$ (containing a small amount of tetramethylsilane, TMS, standard). The solution should fill the lower 4 cm of the NMR tube (using less CDCl$_3$ will lead to problems shimming the magnet). Record the spectrum from $\delta +10$ to $\delta -5$ (relative to TMS at δ 0), expand and plot any complex regions, and integrate the spectrum. Now add about 3 drops of D$_2$O, shake the two phase CDCl$_3$–D$_2$O mixture vigorously for about 15 seconds, and again record the NMR spectrum.

Part B

(Tetraphenylporphyrinato)copper(II), Cu(TPP)

The reaction set up is the same as in Part A (Fig. 23-5). Place 0.1 g (0.16 mmol) of H$_2$TPP in a 100-mL one-neck round-bottom flask along with a stir bar. Add 20 mL of N,N-dimethylformamide (DMF) and stir the mixture (it is not necessary for all of the H$_2$TPP to dissolve). To the dark purple solution, add 0.16 g (0.80 mmol) of hydrated copper(II) acetate, Cu(O$_2$CCH$_3$)$_2 \cdot$ H$_2$O. Fit the flask with a reflux condenser and use grease to lubricate the joint. Bring the reaction mixture to reflux with a rheostat-controlled heating mantle. Allow the reaction to proceed for 30 min.

Because the copper complex is nonfluorescent, conversion to the copper complex can be confirmed by checking for complete quenching of the porphyrin fluorescence under long-wavelength UV light. Spot some of the reaction solution on a non-fluorescing TLC plate with a Pasteur pipet, and examine the plate with a long-wavelength UV light. If the conversion is not complete (i.e., if the spot still glows red—even around the edges), add additional copper(II) acetate to the reaction solution and reflux the mixture for 10 min; then redo the spot test. Alternatively, the conversion to the copper complex can be followed by UV–vis spectroscopy. Dip a capillary tube (open at both ends) into the reaction mixture and then dip the tube containing a little of the reaction solution into a cuvette containing pure dichloromethane. Complete conversion is indicated by the disappearance of the bands at 650 and 592 nm.

When the reaction is complete, cool the reaction mixture to room temperature in an ice–water bath for 5–10 min, and then add 50 mL of distilled water to precipitate the porphyrinic material. Transfer the slurry to a separatory funnel, and extract it three times with 25 mL of dichloromethane. Collect the organic (bottom) layer each time. Discard the aqueous layer that remains in the funnel (it typically is pale blue), and then pour the combined organic extracts back into the separatory funnel. Add 50 mL of distilled water, shake the mixture, and then drain out the organic (bottom) layer. Concentrate the organic layer to dryness on a rotary evaporator. Discard the aqueous wash solution that remains in the separatory funnel.

Chromatographic Purification of Cu(TPP)

In this step, you will use TLC to determine which solvent system will allow you to separate Cu(TPP) from the H_2TPP and chlorin impurities.

With a very small portion of the Cu(TPP) mixture, prepare a concentrated dichloromethane solution for use in the TLC trial separations. A very small amount of this solution will be spotted on the silica gel 8–10 mm from the bottom of the microscope slide. Because the best separations are achieved when this spot is between 3 and 5 mm in diameter, the dichloromethane solution should be applied with a very small capillary. These can be purchased or can be prepared by heating the middle of a melting point capillary tube over a low flame and quickly pulling the ends of the tube apart before the tube is sealed off. Scratch the capillary with a sharp file and break the tube into two applicators. Dip an applicator into the dichloromethane solution of the mixture and touch it to the TLC plate, giving a spot that is not larger than 4 mm in diameter. Allow the dichloromethane to evaporate, and then make a second application of the solution to the same spot.

Fill five screw-cap bottles (Fig. 23-3) with the trial solvents listed below to a depth of 4 mm or less, such that the level is below your spot when the TLC plate is inserted.

1. Hexane (or petroleum ether, 60–70 °C boiling fraction)
2. Toluene
3. Ethyl acetate
4. Dichloromethane
5. Acetone

Insert a spotted TLC plate into each bottle and replace the cover. Allow the solvent to rise about three-fourths of the way up the plate and then remove it and allow it to dry. Do this for each of the five TLC plates, carefully recording which solvent was used for each slide. The Cu(TPP) and the impurities are intensely colored and easily visible even at low concentrations on the plate. In other cases, however, where the compounds to be separated are weakly colored and are difficult to see, the locations of the spots can be determined either chemically (by exposing the plate to a developer such as iodine) or by examining the plate under UV light.

Make a drawing in your research notebook of the location of the spots for each of the five attempts. To establish which spot is H_2TPP, obtain a fresh TLC plate, spotting it first with the dichloromethane solution of the mixture and then in an adjacent position with a dichloromethane solution of pure H_2TPP. Develop the plate in one of the solvents that gave a good separation of spots, and establish which spot of the mixture is H_2TPP by comparing it with the known H_2TPP spot.

From the five TLC trials, you should select a solvent for the column chromatographic separation. You may select a solvent in which one of the components moves rapidly and the other more slowly. Such a solvent should give a good separation on the column. Alternatively, one might choose an initial solvent in which only one component moves while the other remains at the starting point. The first component could then be washed off the column, and then a more polar solvent could be added to elute the other component. Regardless of your choice, it should be based on the separation achieved on the TLC plates.

Clamp a 3-cm diameter, 30-cm long chromatography column to a ring stand or other rigid support. Push a 2-cm diameter wad of glass wool to the bottom of the tube with a rod. Then pour in enough sand to give a 1-cm layer, and add enough solvent to cover the sand. In a medium sized beaker, prepare a slurry of silica gel in your selected solvent. Use enough silica gel to fill the column to a height of about 8–10 cm. Pour the slurry gently into the column and let the excess solvent drain until the meniscus is about 0.5 cm above the settled gel. (Do not allow the solvent level to drop below the top of the gel.) Carefully pour a slurry of silica gel and the Cu(TPP) mixture in a few milliliters of the initial solvent onto the column. (Depending on which solvent you choose, the sample may or may not dissolve completely. Even if it does not dissolve completely, a good separation should result if you have selected your solvent correctly.) Lower the solvent again at the top of the adsorbent and add another 1-cm layer of sand to prevent the bed from being disturbed when the eluting solvent is added. Then gently fill the column with the initial eluting solvent, being careful not to agitate the bed. Carry out the elution, using a flow rate of a few drops per second. Insoluble residues, if any, will remain on top of the column. Generally, two bands move down the column: a slowly moving sharp band that is composed of impurities, and a quickly moving broad red smear that contains the desired Cu(TPP) product. Collect this broad band continuously until all the red material has been eluted, or until the sharp band is about 2 cm from the bottom of the column. Collect only the broad red band due to Cu(TPP).

Concentrate the Cu(TPP) solution to 50 mL on a rotary evaporator and then add 50 mL of methanol to precipitate the product. Filter the solution using a medium glass frit, rinse the product with two 10-mL portions of methanol, and dry the product under vacuum if possible. Record the weight. The used silica gel can be removed by inverting the column over a large-mouth jar. The tar that remains at the top of the column should be discarded along with the used silica gel; be sure to put them into the appropriate waste container.

Obtain the visible (or UV–vis) spectrum of a very dilute toluene solution of the purified Cu(TPP) product (the molar absorption coefficient in benzene is reported to be 20,600 at 538 nm).

REPORT

Include the following:
1. Yields of H_2TPP and Cu(TPP).
2. The ^1H NMR spectra of H_2TPP both with and without added D_2O.
3. Visible (or UV–vis) spectra of H_2TPP and Cu(TPP).

PROBLEMS

1. What are the assignments of the NMR peaks for H_2TPP? Why are some of the peaks multiplets while others are singlets? Why are some of the peaks downfield of TMS (positive shift) and some upfield (negative shift)?

2. How does the NMR spectrum of H_2TPP change upon addition of D_2O? Why are only some of the peaks affected by D_2O?

3. What are the electronic transitions that are responsible for the peaks in the visible spectra of H_2TPP and $Cu(TPP)$, and why are the two spectra different?

4. Why does $Cu(TPP)$ elute before H_2TPP?

5. Why should the reaction of benzaldehyde and pyrrole *not* be carried out under a N_2 atmosphere?

6. In the purification of $Cu(TPP)$, what might be the tar that remained at the top of the chromatography column?

7. The rates at which compounds elute from a silica gel column depend on any pretreatment of the silica gel. In which case would a compound elute faster: down a column made of silica gel that had previously been heated at 150 °C under vacuum for 8 h, or down a column made of silica gel that had been sitting open in the laboratory for a few days? Explain.

8. What methods might be used to detect the elution of colorless compounds from a column?

9. A mixture of cis and trans isomers of the neutral complex $Cr(CO)_4[P(C_6H_5)_3]_2$ is loaded onto a silica gel column and eluted with $CHCl_3$. Which isomer would elute first, and why?

INDEPENDENT STUDIES

A. Prepare and characterize the iron porphyrin complex $Fe(TPP)Cl$ and the imidazole adduct $[Fe(TPP)(Im)_2]Cl$, which are related to the oxidized form of the oxygen-carrying molecule hemoglobin. Use UV–vis to show isosbestic behavior in the reaction of $Fe(TPP)Cl$ with imidazole. (Epstein, L. M.; Straub, D. K.; Maricondi, C. *Inorg. Chem.* **1967**, *6*, 1721; Collins, D. M.; Countryman, R.; Hoard, J. L. *J. Am. Chem. Soc.* **1972**, *94*, 2066.)

B. Brominate $Cu(TPP)$ at all eight pyrrole positions. (Bhyrappa, P.; Krishnam, V. *Inorg. Chem.* **1991**, *30*, 239.)

C. Prepare and characterize the nickel complex $Ni(TPP)$, which is a model of the nickel-containing enzyme called F-430. (Johnson, E. C.; Dolphin, D. *Inorg. Synth.* **1980**, *20*, 143.)

D. Prepare the water soluble porphyrins tetrakis(carboxymethylphenyl)-porphyrin or tetrakis(alkylpyridyl)porphyrin. (Datta-Gupta, N.; Bardos, T. J. *J. Heterocycl. Chem.* **1966**, *3*, 495; Beckmann, B. A.; Bochman, A.; Pasternack, R. F.; Reinprecht, J. T.; Vogel, G. C. *J. Chem. Educ.* **1976**, *53*, 387.)

E. Investigate the use of metals as templates for the synthesis of large-ring compounds such as $[Ni(R_4[14]\text{-}1,3,8,10\text{-tetraeneN}_4)]^{+/0}$. (Tait, A. M.; Busch, D. H. *Inorg. Synth.* **1978**, *18*, 22; Hayes, J. W. II; Taylor, C. J.; Hotz, R. P. *J. Chem. Educ.* **1996**, *73*, 991.)

F. Prepare Goedken's macrocycle and investigate its coordination chemistry. (Chipperfield, J. R.; Woodward, S. *J. Chem. Educ.* **1994**, *71*, 75.)

G. Study the uptake of O_2 by a cobalt(II) Schiff-base complex. (Aymes, D. J.; Paris, M. R. *J. Chem. Educ.* **1989**, *66*, 854.)

H. Use microscale techniques to prepare H_2TPP, Zn(TPP), and Ni(TPP). (Marsh, D. F.; Mink, L. M. *J. Chem. Educ* **1996**, *73*, 1188.)

I. Record and interpret the EPR spectrum of Cu(TPP). (For a recent EPR study of a copper(II) complex, see Louloudi, M.; Deligiannakis, Y.; Tuchagues, J.-P.; Donnadieu, B.; Hadjiliadis, N. *Inorg. Chem.* **1997**, *36*, 6335.)

REFERENCES

Porphyrins and Metalloporphyrins

Adler, A. D.; Longo, F. R.; Finarelli, J. D.; Goldmacher, J.; Assour, J.; Korsakoff, L. *J. Org. Chem.* **1967**, *32,* 476. Synthesis of H_2TPP.

Adler, A. D.; Longo, F. R.; Váradi, V. *Inorg. Synth.* **1976**, *16*, 213. Synthesis of H_2TPP.

Abraham, R. J.; Jackson, A. H.; Kenner, G. W.; Warburtion, D. *J. Chem. Soc.* **1963**, 853. NMR spectrum of H_2TPP.

Collman, J. P.; Wagenknecht, P. S.; Hutchison, J. E. *Angew. Chem., Int. Ed. Engl.* **1994**, *33*, 1537. A review of face-to-face porphyrin complexes.

Dolphin, D. *The Porphyrins*, Academic: New York, 1979. An excellent multivolume review of porphyrin chemistry.

Lindsey, J. S.; Schreiman, I. C.; Hsu, H. C.; Kearney, P. C.; Marguerettaz, A. M. *J. Org. Chem.* **1987**, *52*, 827. Alternative synthesis of H_2TPP and other porphyrin rings.

Scheidt, W. R.; Reed, C. A. *Chem. Rev.* **1981**, *81*, 543. Review of iron porphyrin chemistry.

Mashiko, T.; Dolphin, D. in *Comprehensive Coordination Chemistry*, Wilkinson, G.; Gillard, R. D.; McCleverty, J. A., Eds, Pergamon: New York, 1987; Chapter 21. Excellent review of metallaporphyrins and related complexes of phthalocyanines and other macrocyclic nitrogenous ligands.

Chromatographic Techniques

Braithwaite, A.; Smith, F. J. *Chromatographic Methods*, 4th ed., Chapman and Hall: New York, 1985.

Gritter, R. J.; Bobbitt, J. M.; Schwarting, A. E. *Introduction to Chromatography*, 2nd ed., Holden-Day: Oakland, CA, 1985. An excellent, practical introduction to thin layer, column, and gas chromatography. Contains extensive general references and a list of chromatographic equipment suppliers.

Druding, L. F.; Kauffman, G. B. *Coord. Chem. Rev.* **1968**, *3*, 409. Thin layer, column, and paper chromatography of coordination complexes.

Guiochon, G.; Pommier, C. *Gas Chromatography in Inorganics and Organometallics*, Ann Arbor Science: Ann Arbor, MI, 1973. Excellent coverage of experimental techniques and the older literature.

Miller, J. M. *Chromatography: Concepts and Contrasts*, Wiley: New York, 1988.

Poole, C. F.; Poole, S. K. *Chromatography Today*, Elsevier: New York, 1991.

For General References to Bioinorganic Chemistry, see Experiment 21

OVERVIEW

In a typical 14-week class, our students conduct between six and eight "standard" experiments. It is convenient to start the class with one of the easier preparations [e.g. Cu(TPP), Co(en)$_3^{3+}$, VO(PO$_4$)$\cdot$2H$_2$O] that can be performed on the bench-top and that require only routine equipment. Such an experiment can be started with no advance preparation on the students' part and it instructs them in many routine operations such as weighing, use of reflux condensers, use of magnetic stirring plates, and so on. After this experiment, the students conduct their remaining experiments according to a preset schedule. These experiments are staggered in order to allow the efficient use of equipment such as the tube furnace and vacuum line. The final 2 or 3 weeks are dedicated to a "special project," which can be selected from the Independent Study sections at the end of each chapter.

We placed the equipment for each experiment in locked drawers in the laboratory. Because several students will use the same equipment throughout the course, it is essential that glassware be left clean and in good condition at the end of each laboratory period. Additionally, students were assigned their own drawers in which they could store a minimum of equipment, such as spatulas, NMR tubes, stirring bars, or compounds they had prepared.

There are periods during some experiments when the student has little to do. We encouraged our students to use this time to complete another experiment. With careful planning, it is possible for the students to conduct two experiments simultaneously such that the waiting period for one experiment [e.g., VO(PO$_4$)$\cdot$2H$_2$O] coincides with the active stage of a second experiment (YBa$_2$Cu$_3$O$_7$).

Notes on Experiments

The indicated sources of supplies and equipment are simply those that we have found satisfactory. That we have included specific brand names in no way implies that these items are superior to others on the market.

Waste diposal is an important consideration. We provided separate polyethylene bottles for the wastes for each of the experiments. Mercury wastes were kept separate.

Experiment 1. The Standard Temperature Tube Furnace from Cole Parmer (7425 North Oak Park Avenue, IL 60714) is adequate. The fused silica tube was 5 cm in diameter and 50 cm long. Fused silica tubing can be purchased from Fredrick and Demmock (P.O. Box 230, Millville, NJ 08332). Thermocouples were obtained from Omega Engineering (P.O. Box 4047, Stamford, CT 06907). If the thermocouple is used to control the heating, make certain that it is *in* the tube furnace before turning the furnace on! The oxygen flow should also be established before turning on the furnace. After many uses the silica tube will devitrify (crystallize) as indicated by the development of an opaque white appearance. Such tubes need to be replaced because they can be very brittle.

We allowed only one sample in the tube furnace at any time. Typically, pairs of students did the high-temperature synthesis jointly but conducted their titrations separately. We stored $BaCO_3$ on the shelf, but other materials were stored in a desiccator. The weighing and grinding together of the powders takes up to an hour and should be started as soon as possible in the lab period. For the iodometry, we used a premade potassium iodate standard solution, for example, from Aldrich Chemical Co. (1001 W. St. Paul Ave., Milwaukee, WI 53233). Students will need to be reminded that they will weigh the boat a total of three times: tare weight, before reaction, and after reaction (the idea of determining a product formula gravimetrically will be fairly new to them). Cleaning the reactants and the superconductor from the mortar and pestle is easily accomplished with 6 M HCl. Cleaning the boat may require hot concentrated HCl. The X-ray powder pattern of the product may differ from that reported in Appendix 10 owing to impurities or preferential orientation effects.

Experiment 2. The compound $VO(PO_4) \cdot 2H_2O$ is easily reduced as indicated by the development of a blue-green tinge on the product, which should not affect the X-ray diffraction analysis or the subsequent reduction step.

Experiment 3. The silica gel and aluminum isopropoxide were obtained from Aldrich. The silica gel was TLC standard grade without gypsum binder or fluorescent indicator, 2–25 μm particle size, and the aluminum isopropoxide was the 9890 grade. We have routinely used the Parr bomb shown in Figure 3-2. It consists of a Model 4753 general purpose stainless steel bomb (4 × 18 cm, 200-mL capacity), a Model A427HC gauge block assembly with silver rupture disk, and Model A495HC hose assembly. It is available from Parr Instrument Company (211 53rd St., Moline, Il 61265). Do not use a glass liner.

Workers at high altitudes may need to compensate for the lower boiling point of water.

Experiment 4. The $NaBH_4$ needs to be fairly fresh, as this salt has a limited shelf life.

Experiment 5. We purchased fullerene soot from Southern Chemical Group (2155 W. Park Court, Suite A, Stone Mountain, GA 30087) and from SES Research (6008 W. 34th, Suite H, Houston, TX 77092). For cyclic voltammetry, we used a CV-50W voltammetric analyzer (Bioanalytical Systems, 2701 Kent Avenue, Purdue Research Park, West Lafayette, IN 47906). The $N(C_4H_9)_4PF_6$ is a generally useful electrolyte that is nonhygroscopic, soluble in many solvents, and readily purified. The preparation (from $N(C_4H_9)_4I$ and NH_4PF_6) and purification of this salt (and other reagents and solvents used in electrochemistry) are described in Fry, A. J.; Britton, W. E. in *Laboratory Techniques in Electro-analytical Chemistry*; Kissinger, P. T.; Heineman, W. R., Eds., Marcel Dekker: New York, 1984, p 367.

The slow step in this experiment is the evaporation of the *o*-dichlorobenzene after the chromatography. For this purpose, we used a vacuum pump (protected with a trap cooled with liquid N_2).

Experiment 6. Although this synthesis is very reliable, occasional problems arise due to clogging of the U-tubes. This problem can be minimized by using U-tubes with larger bulbs. On our utility vacuum line, we have used 14/20 standard taper joints everywhere except on the cold trap, where a 45/50 joint is used. The bulb used to determine the vapor pressure of germane had a volume of about 300 cm^3.

High-vacuum stopcocks were obtained from ChemGlass (3861 N. Mill Rd., Vineland, NJ 08360), for glass stopcocks, and Kimble-Kontes (1022 Spruce St., Vineland, NJ 08360), for Teflon stopcocks. Be sure to clamp the ball-and-socket joints. Premade vacuum lines can be purchased from O'Brien Scientific Glassblowing (P.O. Box 495, 725A W. Bridge St., Monticello, IL 61856).

Experiment 7. Sometimes the product does not crystallize from the ethyl acetate in which case the volume of this solution needs to be reduced further. This experiment avoids the use of the highly toxic methyl and ethyl tin compounds.

Experiment 8. We used a cylinder containing 6.8 kg of ammonia. We placed the cylinder on a dolly and elevated the bottom of the cylinder by strapping it to a block of wood.

This experiment is easier than it might appear. An instructor needs to assist in the transferring the ammonia from the cylinder to the flask. We dispensed the ammonia *before* adding the dry ice to the condenser, in order to minimize the condensation of moisture from the atmosphere. The reaction can also conducted without dry ice but it is better in such cases to start with more ammonia. The extraction of the products from the three-neck reaction flask proceeds well: Students should avoid trying to chip out crude product from the bottom of the flask.

Experiment 9. A variety of power supplies can be obtained from Cole Parmer (7425 North Oak Park Avenue, IL 60714). For a convenient method of sealing metal wires into glass tubing, see Vijh, A. K.; Alwitt, R. S. *J. Chem. Educ.* **1969**, *46*, 121. $K_2S_2O_8$ can be purchased if the instructor chooses to focus on the second half of this experiment.

Experiment 10. The used Dowex 50W-X8 resin may be regenerated by passing 6 *M* HCl through it, followed with a water wash until the effluent is no longer acidic.

Experiment 11. We often did only the first two steps in this otherwise long procedure. Tributylphosphine was stored in a Schlenk flask from which the students drew out the amount needed with a syringe.

Experiment 12. We used a magnetic susceptibility balance from John Matthey (436 Devon Park Dr., Wayne, PA 19087). Instructions for the construction of a traditional Gouy balance can found in the second edition of this text. The calibration standards, $HgCo(NCS)_4$ and $[Ni(en)_3]S_2O_3$, and their syntheses are described in the following references: Figgis, B. N.; Nyholm, R. S. *J. Chem. Soc.* **1958**, 4190; Curtis, N. F. *J. Chem. Soc.* **1961**, 3147.

Experiment 13. Students interested in kinetics should only undertake the preparation and aquation of $Co(NH_3)_5Cl^{2+}$. We have used an YSI conductivity bridge, model 4302 (YSI Inc., Yellow Springs, OH 44387). Mineral oil or water covered with a layer of mineral oil has been used for the 60 °C thermostatted bath. The temperature of the bath should be constant within ± 0.2 °C.

Experiment 14. The synthesis has been completely revised since the previous edition and proceeds more reliably and quickly. We used a DIP-360 digital polarimeter purchased from JASCO (8649 Commerce Dr., Easton, MD 21601). The polarimeter cell, also from JASCO, had a 1-decimeter path length.

Experiment 15. The combination of CS_2 and Na might appear to be dangerous due to a possible runaway exothermic reaction. The original report notes no hazards and our own extensive testing have revealed no problems. One can conclude the experiment after the isolation of $[N(C_2H_5)_4]_2[Zn(C_3S_5)_2]$. Alternatively, because $[N(C_2H_5)_4]_2[Zn(C_3S_5)_2]$ can be easily prepared on a 25-g scale, students could be given this salt and start off by preparing the nickel complex.

Experiment 16. Solution IR cells can be purchased from International Crystal Laboratories (11 Erie St., Garfield, NJ 07026).

Experiment 17. The sensitivity of $(C_5H_5)Fe(CO)_2CH_3$ to air makes this synthesis somewhat challenging. Our students usually were provided $(C_5H_5)_2Fe_2(CO)_4$, which is easily prepared in advance on a multigram scale. When subliming the final product, students should submerge as much of the Schlenk flask as possible in the water (50 °C). Schlenk flasks are not necessary for this experiment, but they can be purchased from Kimble-Kontes (1022 Spruce St., Vineland, NJ 08360). The reduction step can also be carried out with sodium amalgam (see the second edition of this text), but disposal of the mercury waste can be a problem. The reduction should *not* be carried out with $LiBH(C_2H_5)_3$ in place of the potassium salt.

Experiment 18. Temperature control is the most important aspect of the first step in this experiment. Attempts to purify $Fe_3(CO)_{12}$ by sublimation or Soxhlet extraction result in substantial decomposition to metallic iron. Although $Fe_3(CO)_{12}$ can be handled in air, it is occasionally pyrophoric, hence the recommendation that it not be dried by pulling a stream of air through it.

Experiment 19. The iridium salt was dispensed in preweighed vials. We purchased iridium salts from Pressure Chemical (3419 Smallman St., Pittsburgh, PA

15201). It is essential to boil the reaction solution vigorously. Reaction times up to 18 h are acceptable (and even beneficial).

Experiment 20. This experiment is fairly advanced. The $[Ni(NH_3)_6]Cl_2$ should be fresh because it loses NH_3 upon standing. This complex can be prepared while the Na reacts with the cyclopentadiene. Cyclopentadiene has an objectionable odor and is toxic, thus the cracking should be conducted in a hood. The "Cp cracker" can be reused for many weeks; any unused cyclopentadiene can be added back to the distillation flask. Do not fill the distillation flask more than about one-third full with dicyclopentadiene: The cracking process is accompanied by foaming, which can carry uncracked dicyclopentadiene into the receiver. The synthesis and purification of nickelocene requires one busy lab period followed by a lab period that is dedicated only to the sublimation step. At the end of the experiment, we often asked the students to flame-seal their samples in ampules on the vacuum line (nickelocene has a short shelf life in air). This assignment gives students some experience with glassblowing.

Experiment 21. This experiment is a useful introduction to the use of inert atmosphere techniques. Good results are obtained even by those with no previous experience with inert atmosphere techniques. We recommend that the solution of Br_2 in acetic acid be prepared in advance by the instructor.

Experiment 22. We used relatively inexpensive pH meters and standard electrodes with good results. The standardizing buffers were purchased.

Experiment 23. This experiment works well despite the low yields ($\sim 10\%$) for the H_2TPP synthesis. Pyrrole oxidizes slowly in air, and better yields of H_2TPP can be obtained if the pyrrole is freshly distilled. The experiment can be shortened by omitting the TLC study of Cu(TPP) and proceeding directly to the chromatographic purification using dichloromethane as the eluting solvent.

Cuvettes for UV–vis spectroscopy can be purchased from many supply houses; we used products from Hellma Cells (Box 544, Borough Hall St., Jamaica, NY 11424). A hand-held 4-watt UV light works well for detecting the free H_2TPP; these can be purchased from Cole Parmer. For column chromatography, we used 0.040–0.063 mm, 230–400 mesh ASTM silica gel (Fisher Scientific, 1600 W. Glenlake, Itasca, IL 60142). For TLC, we used non-fluorescent Silica Gel 60 plates, which were obtained in boxes of 20 × 20 cm sheets with a layer thickness of 0.25 mm (EM Industries, Inc., 480 Democrat Road, Gibbstown, NJ 08027). The sheets were cut with scissors into 1 × 5 cm strips, which were stored in a screw-capped jar.

Appendices

Approximate Concentrations and Densities of Commercial, Reagent Grade Acids and Bases

Acid or Base	% by Weight	Molarity	Density
Acetic acid (glacial)	99.8	17.4	1.05
Ammonia (aqueous)	29	14.8	0.90
Hydrobromic acid	48	8.9	1.50
Hydrochloric acid	37	12.0	1.19
Hydrofluoric acid	49	28.9	1.16
Nitric acid	70	15.9	1.50
Perchloric acid	70	11.7	1.67
Phosphoric acid	85	14.7	1.70
Sulfuric acid	96	18	1.84

Expected Molar Conductance (Λ_M) Ranges[a] for 2, 3, 4, and 5 Ion Electrolytes ($\sim 10^{-3}$ M) in Some Common Solvents at 25 °C[b]

Solvent	Dielectric Constant	Electrolyte Types			
		1:1	2:1	3:1	4:1
Water	78.4	118–131	235–273	408–435	~560
Nitromethane	35.9	75–95	150–180	220–260	290–330
Nitrobenzene	34.8	20–30	50–60	70–82	90–100
Acetone	20.7	100–140	160–200		
Acetonitrile	36.2	120–160	220–300	340–420	
N,N-Dimethylformamide	36.7	65–90	130–170	200–240	
Methanol	32.6	80–115	160–220		
Ethanol	24.3	35–45	70–90		

[a] Units on all molar conductances are ohm^{-1} cm^2 mol^{-1}.
[b] Geary, W. J. *Coord. Chem. Rev.* **1971**, *7*, 81.

Transmission Ranges of Cell Materials for Infrared and Ultraviolet–Visible Spectroscopy

	Infrared Region[a]	
Material	Useful Transmission Range (cm^{-1})	Comments
NaCl	5000–625	Low cost, fogs slowly, but is water soluble
KBr	5000–400	Properties similar to NaCl
CaF$_2$	5000–1110	Insoluble in water, resists many acids and bases, difficult to polish
BaF$_2$	5000–830	Properties similar to CaF$_2$
CsBr	5000–250	Water soluble and very hygroscopic, difficult to polish
CsI	5000–165	Properties similar to CsBr, but less hygroscopic
AgCl	5000–435	Insoluble in water, darkens in light, difficult to polish
AgBr	5000–285	Properties similar to AgCl
KRS-5 (TlBr, TlI)	5000–250	Slightly soluble in water, toxic
Irtran-2 (ZnS)	5000–835	Insoluble in water, rugged
Sapphire (Al$_2$O$_3$)	5000–780	Insoluble in water, hard
Quartz (SiO$_2$)	5000–2700	Insoluble in water, rugged, limited frequency range
Ge	5000–600	Insoluble in water, brittle, high reflection losses
Si	5000–660	Properties similar to Ge
Polyethylene	625–33	Insoluble in water, low cost, difficult to clean
	Ultraviolet–Visible Region (nm)	
Pyrex	320–2500	
Vycor	320–2500	
Standard Silica	220–2600	
NIR Silica	220–3500	
Far-UV Silica	160–2600	

[a] Miller, R. G. J., Ed., *Laboratory Methods in Infrared Spectroscopy*, 2nd ed., Heyden: London, 1972.

Properties of Commercial Glasses[a]

	Soda-Lime (Soft; Flint; TEKK; EXAX)	Borosilicate (Kimax; Pyrex)	96% Silica (Vycor)	Fused Silica
Strain point[b] (°C)	480	510	820	1020
Annealing point[c] (°C)	520	550	910	1120
Softening point[d] (°C)	700	820	1500	1710
Working point[e] (°C)	1010	1245		
Linear coefficient of expansion[f]	9.3	3.3	0.8	0.6
Refractive index (n_D)	1.52	1.474	1.458	1.54
Density (g cm^{-3})	2.5	2.23	2.18	2.65

[a] Gordon, A. J.; Ford, R. A. *The Chemist's Companion*, Wiley: New York, 1972.

[b] Point at which strain is relieved after 4 h.

[c] Point at which strain is relieved in 15 min.

[d] Point at which sagging begins.

[e] Point at which glass can be easily manipulated and blown by mouth.

[f] Parts per million per degree Celcius (°C) between 0 and 300 °C.

Common Solvents and Their Properties

Name, Formula	bp^a (°C)	mp^a (°C)	ε^b
Acetic acid, CH_3CO_2H	118	16.6	6.2
Acetone, $(CH_3)_2CO$	56.2	−95.4	20.7
Acetonitrile, CH_3CN	81.6	−45.7	36.2
Ammonia, NH_3	−33.4	−77.7	27 (−60)
Benzene, C_6H_6	80.1	5.5	2.3 (20)
Carbon disulfide, CS_2	46.2	−111.5	2.6
Carbon tetrachloride, CCl_4	76.5	−23	2.2
Chlorobenzene, C_6H_5Cl	132	−46	5.6
Chloroform, $CHCl_3$	61.7	−63.5	4.7
Cyclohexane, $(CH_2)_6$	80.7	6.5	2.0
Decalin, $C_{10}H_{18}$	189	−125	2.2 (20)
1,2-Dichloroethane, $ClCH_2CH_2Cl$	83.5	−35	10.4
o-Dichlorobenzene, $C_6H_4Cl_2$	180	−17	9.9
Dichloromethane, CH_2Cl_2	40	−95	8.9
Diethyl ether, $(C_2H_5)_2O$	34.5	−116	4.3 (20)
Diglyme, $(CH_3OCH_2CH_2)_2O$	162	−64	7.2
Dimethoxyethane (Glyme), $CH_3OCH_2CH_2OCH_3$	83	−58	7.3
N,N-Dimethylformamide, $HC(O)N(CH_3)_2$	152	−61	36.7
Dimethyl sulfoxide, $(CH_3)_2SO$	189	18.4	49
1,4-Dioxane, $O(CH_2CH_2)_2O$	102	11.8	2.2
Ethanol, CH_3CH_2OH	78.3	−114	24.3
Ethyl acetate, $CH_3CO_2C_2H_5$	77.1	−83.6	6.0
Hexamethylphosphoramide, $[(CH_3)_2N]_3PO$	233	7.2	30 (20)
n-Hexane, $CH_3(CH_2)_4CH_3$	69	−95	1.9 (20)
Hydrogen cyanide, HCN	26	−14	115 (20)
Hydrogen fluoride, HF	19.5	−89.4	84 (0)
Methanol, CH_3OH	64.5	−97.5	32.6
Mesitylene, $C_6H_3(CH_3)_3$	164.7	−44.7	2.3 (20)
Nitrobenzene, $C_6H_5NO_2$	211	5.8	35 (30)
Nitromethane, CH_3NO_2	101	−28.5	38.6
n-Pentane, $CH_3(CH_2)_3CH_3$	36.1	−130	1.8
i-Propanol, $(CH_3)_2CHOH$	82.4	−89.5	18.3
Pyridine, C_5H_5N	116	−42	12.3
Sulfur dioxide, SO_2	−10.1	−75.5	15.4 (0)
Tetrahydrofuran, $(CH_2)_4O$	66	−65	7.3
Thionyl chloride, $SOCl_2$	80	−105	9.2 (20)
Toluene, $C_6H_5CH_3$	111	−95	2.4
Water, H_2O	100.0	0	78.5

^{a} Boiling points and melting points are at 760 Torr (1 atm).
^{b} Dielectric constants are at 25 °C unless indicated otherwise in parentheses.

APPENDIX 6

Drying (Dehydrating) Agents[a]

Agent	Properties
$Al_2O_3^{b}$ or SiO_2	Very high capacity and fast; good for solvents or gases; may be regenerated at high temperature under vacuum. Equilibrium water vapor pressure above them is about 10^{-3} Torr.
BaO or CaO	Forms $Ba(OH)_2$ or $Ca(OH)_2$; cannot be used with solvents sensitive to base. Slow but efficient. Water vapor pressures above BaO and CaO are 7×10^{-4} and 3×10^{-3} Torr, respectively.
$CaCl_2$	Very fast, but not efficient (0.2 Torr water vapor pressure).
CaH_2	Forms H_2 and $Ca(OH)_2$. Excellent efficiency ($< 10^{-5}$ Torr water vapor pressure). Do not use with halogenated solvents or those with active groups such as aldehydes. Slower but just as efficient as $LiAlH_4$.
$CaSO_4$	Low capacity but fast. Water vapor pressure above it is 5×10^{-3} Torr. Can be regenerated. Sold commercially as "Drierite."
H_2SO_4 (concd)	Fast and has high capacity. Water vapor pressure is about 3×10^{-3} Torr. Cannot be used with basic compounds.
KOH	Fast and very high capacity. Water vapor pressure of about 2×10^{-3} Torr. Especially good for amines, pyridine.
$LiAlH_4$	Forms H_2, LiOH, and $Al(OH)_3$. Use only with inert solvents, because it reacts with a variety of organic functional groups. Can decompose explosively if heated too strongly.
$MgSO_4$	Fast and fair capacity; an excellent general drying agent.
Molecular sieves (Types 3A and 4A)	Fast and high capacity. Very efficient (about 1×10^{-3} Torr water vapor pressure). May be regenerated at 350 °C in vacuum.
Na	Forms H_2 and NaOH. Use only with inert solvents; explosive with halogenated organic solvents. Commonly used with benzophenone to generate deep blue solutions of $Na^+(C_6H_5)_2CO^-$.
Na_2SO_4	Slow and inefficient, but high capacity. Good for predrying. May be regenerated at 150 °C.
P_4O_{10}	Forms phosphoric acid. Very fast and efficient (water vapor pressure of 2×10^{-5} Torr).

[a] Gordon, A. J.; Ford, R. A. *The Chemist's Companion*, Wiley: New York, 1972, p 445. Shriver, D. F.; Drezdzon, M. A. *The Manipulation of Air-Sensitive Compounds*, 2nd ed., Wiley: New York, 1986. Armarego, W. L. F.; Perrin, D. D. *Purification of Laboratory Chemicals*, 4th ed., Butterworth–Heinemann: Oxford, UK, 1996.
[b] A solvent purification system based on filtration through activated Al_2O_3 has been developed: Pangborn, A. B.; Giardello, M. A.; Grubbs, R. H.; Rosen, R. K.; Timmers, F. J. *Organometallics* **1996**, *15*, 1518.

Natural Abundances and Nuclear Spins of Stable Isotopes[a]

Isotope	Natural Abundance (%)	Nuclear Spin[c,d] (I)	Isotope	Natural Abundance (%)	Nuclear Spin[c,d] (I)
^{1}H	99.985	1/2	^{39}K	93.22	3/2
^{2}H	0.015	1	^{41}K	6.77	3/2
^{4}He	100		^{40}Ca	96.97	
^{6}Li	7.42	1	^{42}Ca	0.64	
^{7}Li	92.58	3/2	^{44}Ca	2.06	
^{9}Be	100	3/2	^{45}Sc	100	7/2
^{10}B	19.7	3	^{46}Ti	7.99	
^{11}B	80.3	3/2	^{47}Ti	7.32	5/2
^{12}C	98.892		^{48}Ti	73.99	
^{13}C	1.108	1/2	^{49}Ti	5.46	7/2
^{14}N	99.635	1	^{50}Ti	5.25	
^{15}N	0.365	1/2	^{50}V	0.25	6
^{16}O	99.759		^{51}V	99.75	7/2
^{17}O	0.037	5/2	^{50}Cr	4.31	
^{18}O	0.204		^{52}Cr	83.76	
^{19}F	100	1/2	^{53}Cr	9.55	3/2
^{20}Ne	90.92		^{54}Cr	2.38	
^{21}Ne	0.257	3/2	^{55}Mn	100	5/2
^{22}Ne	8.82		^{54}Fe	5.84	
^{23}Na	100	3/2	^{56}Fe	91.68	
^{24}Mg	78.60		^{57}Fe	2.17	1/2
^{25}Mg	10.11	5/2	^{58}Fe	0.31	
^{26}Mg	11.29		^{59}Co	100	7/2
^{27}Al	100	5/2	^{58}Ni	67.76	
^{28}Si	92.18		^{60}Ni	26.16	
^{29}Si	4.71	1/2	^{61}Ni	1.25	3/2
^{30}Si	3.12		^{62}Ni	3.66	
^{31}P	100	1/2	^{64}Ni	1.16	
^{32}S	95.0		^{63}Cu	69.1	3/2
^{33}S	0.76	3/2	^{65}Cu	30.9	3/2
^{34}S	4.22		^{64}Zn	48.89	
^{35}Cl	75.53	3/2	^{66}Zn	27.81	
^{37}Cl	24.47	3/2	^{67}Zn	4.11	5/2
^{40}Ar	99.60[b]		^{68}Zn	18.56	

Appendix 7

Natural Abundances and Nuclear Spins of Stable Isotopes[a] (*continued*)

Isotope	Natural Abundance (%)	Nuclear Spin[c,d] (I)	Isotope	Natural Abundance (%)	Nuclear Spin[c,d] (I)
^{70}Zn	0.62		^{101}Ru	17.02	5/2
^{69}Ga	60.2	3/2	^{102}Ru	31.6	
^{71}Ga	39.8	3/2	^{104}Ru	18.87	
^{70}Ge	20.55		^{103}Rh	100	1/2
^{72}Ge	27.37		^{102}Pd	0.96	
^{73}Ge	7.67	9/2	^{104}Pd	10.97	
^{74}Ge	36.74		^{105}Pd	22.2	5/2
^{76}Ge	7.67		^{106}Pd	27.3	
^{75}As	100	3/2	^{108}Pd	26.7	
^{74}Se	0.87		^{110}Pd	11.8	
^{76}Se	9.02		^{107}Ag	51.35	1/2
^{77}Se	7.58	1/2	^{109}Ag	48.65	1/2
^{78}Se	23.52		^{106}Cd	1.22	
^{80}Se	49.82		^{108}Cd	0.88	
^{82}Se	9.19		^{110}Cd	12.39	
^{79}Br	50.52	3/2	^{111}Cd	12.75	1/2
^{81}Br	49.48	3/2	^{112}Cd	24.07	
^{84}Kr	56.90[b]		^{113}Cd	12.26	1/2
^{85}Rb	72.15	5/2	^{114}Cd	28.86	
^{87}Rb	27.85	3/2	^{116}Cd	7.58	
^{84}Sr	0.56		^{113}In	4.23	9/2
^{86}Sr	9.86		^{115}In	95.77	9/2
^{87}Sr	7.02	9/2	^{112}Sn	0.95	
^{88}Sr	82.56		^{114}Sn	0.65	
^{89}Y	100	1/2	^{115}Sn	0.34	1/2
^{90}Zr	51.46		^{116}Sn	14.24	
^{91}Zr	11.23	5/2	^{117}Sn	7.57	1/2
^{92}Zr	17.11		^{118}Sn	24.01	
^{94}Zr	17.40		^{119}Sn	8.58	1/2
^{96}Zr	2.80		^{120}Sn	32.97	
^{93}Nb	100	9/2	^{122}Sn	4.71	
^{92}Mo	15.86		^{124}Sn	5.98	
^{94}Mo	9.12		^{121}Sb	57.25	5/2
^{95}Mo	15.70	5/2	^{123}Sb	42.75	7/2
^{96}Mo	16.50		^{122}Te	2.46	
^{97}Mo	9.45	5/2	^{123}Te	0.87	1/2
^{98}Mo	23.75		^{124}Te	4.61	
^{100}Mo	9.62		^{125}Te	6.99	1/2
^{96}Ru	5.46		^{126}Te	18.71	
^{98}Ru	1.87		^{128}Te	31.79	
^{99}Ru	12.63	5/2	^{130}Te	34.49	
^{100}Ru	12.53		^{127}I	100	5/2

Natural Abundances and Nuclear Spins of Stable Isotopes[a] (*continued*)

Isotope	Natural Abundance (%)	Nuclear Spin[c,d] (I)	Isotope	Natural Abundance (%)	Nuclear Spin[c,d] (I)
^{129}Xe	26.44[b]	1/2	^{178}Hf	27.1	
^{132}Xe	26.89		^{179}Hf	13.75	9/2
^{133}Cs	100	7/2	^{180}Hf	35.22	
^{134}Ba	2.42		^{181}Ta	99.99	7/2
^{135}Ba	6.59	3/2	^{182}W	26.4	
^{136}Ba	7.81		^{183}W	14.4	1/2
^{137}Ba	11.32	3/2	^{184}W	30.6	
^{138}Ba	71.66		^{186}W	28.4	
^{139}La	99.91	7/2	^{185}Re	37.07	5/2
^{138}Ce	0.25		^{187}Re	62.93	5/2
^{140}Ce	88.48		^{186}Os	1.59	
^{142}Ce	11.07		^{187}Os	1.64	1/2
^{141}Pr	100	5/2	^{188}Os	13.3	
^{142}Nd	27.13[b]		^{189}Os	16.1	3/2
^{144}Nd	23.87		^{190}Os	26.4	
^{152}Sm	26.63[b]		^{192}Os	41.0	
^{154}Sm	22.53		^{191}Ir	38.5	3/2
^{151}Eu	47.77	5/2	^{193}Ir	61.5	3/2
^{153}Eu	52.23	5/2	^{192}Pt	0.78	
^{156}Gd	20.47[b]		^{194}Pt	32.9	
^{158}Gd	24.87		^{195}Pt	33.8	1/2
^{160}Gd	21.9		^{196}Pt	25.2	
^{159}Td	100	3/2	^{198}Pt	7.19	
^{162}Dy	25.53[b]		^{197}Au	100	3/2
^{163}Dy	24.97	7/2	^{198}Hg	10.02	
^{164}Dy	28.18		^{199}Hg	16.84	1/2
^{165}Ho	100	7/2	^{200}Hg	23.13	
^{166}Er	33.41[b]		^{201}Hg	13.22	3/2
^{167}Er	22.94	7/2	^{202}Hg	29.80	
^{168}Er	27.07		^{204}Hg	6.85	
^{169}Tm	100	1/2	^{203}Tl	29.50	1/2
^{172}Yb	21.82[b]		^{205}Tl	70.50	1/2
^{174}Yb	31.84		^{204}Pb	1.40	
^{175}Lu	97.40	7/2	^{206}Pb	25.1	
^{176}Lu	2.60	7	^{207}Pb	21.7	1/2
^{176}Hf	5.21		^{208}Pb	52.3	
^{177}Hf	18.56	7/2	^{209}Bi	100	9/2

[a] Lide, D. R., Ed., *CRC Handbook of Chemistry and Physics*, 75th ed., CRC Press, Boca Raton, FL, 1994.

[b] Other isotopes not listed here.

[c] In multiples of $h/2\pi$.

[d] Brevard, C.; Granger, P. *Handbook of High Resolution Multinuclear NMR*, Wiley: New York, 1981.

Baths for Cooling and Heating

I. Cold Baths[a,b]

Solvent	Bath Temperature (°C)
Carbon tetrachloride[a,b]	−23
Acetonitrile[a]	−42
Chlorobenzene[a]	−45
Cyclohexanone[a]	−46
Chloroform[a,b]	−61
Acetone[a]	−78
Ethyl acetate[b]	−84
Toluene[b]	−95
Methanol[b]	−98
Cyclohexene[b]	−104
Carbon disulfide[b]	−112
Ethyl alcohol[b]	−116
n-Pentane[b]	−130
Isopentane[b]	−160

[a] Small lumps of solid CO_2 are added to the solvent until a small excess of the solid is present. Phipps. A. M.; Hume, D. N. *J. Chem. Educ.* **1968**, *45*, 664. Baths useful in the −50 to 0 °C temperature range have been prepared by adding solid CO_2 to various concentrations of $CaCl_2–H_2O$ solutions. Bryan, W. P.; Byrne, R. H. *J. Chem. Educ.* **1970**, *47*, 361.

[b] Liquid N_2 is slowly poured into a Dewar flask containing the organic solvent. The mixture is stirred continuously during the addition to avoid forming a crust on the surface that could break the Dewar. Rondeau, R. E. *J. Chem. Eng. Data* **1966**, *11*, 124.

II. Heating Baths

Bath Material	mp (°C)	Max Safe Working T (°C)
Water	0	100
Mineral oil		180
Silicone oils[a]		250
Dibutyl phthalate		340
Wood's metal[b]	70	500

[a] Very stable, but moderately expensive. Available in several fractions with useful ranges from 30 to 280 °C from Dow Corning.

[b] 50% Bi, 25% Pb, 12.5% Sn, 12.5% Cd.

SI Units

The International Bureau of Weights and Measures has recommended an international system (*Système International d'Unites*) for expressing quantitative units in a consistent and logical manner. It consists of several basic units (length, mass, time, etc.) and many derived units (volume, velocity, force, energy, etc.). Below are listed some basic units, derived units, and prefixes in the *Système International d'Unites*. For a more complete description, see the references.

I. Basic SI Units

Physical Quantity	Name of Unit	Symbol
Length	meter	m
Mass	kilogram	kg
Time	second	s
Electric current	ampere	A
Thermodynamic temperature	degree kelvin	K
Amount of substance	mole	mol

II. Some Derived SI Units

Physical Quantity	SI Name	SI Symbol
Area	square meter	m^2
Volume	cubic meter	m^3
Density	kilogram per cubic meter	$kg\ m^{-3}$
Velocity	meter per second	$m\ s^{-1}$
Acceleration	meter per second squared	$m\ s^{-2}$
Force	newton (N)	$kg\ m\ s^{-2} = J\ m^{-1}$
Pressure	newton per square meter	$N\ m^2$
Energy	joule (J)	$kg\ m^2\ s^{-2} = N\ m$
Power	watt (W)	$kg\ m^2\ s^{-3} = J\ s^{-1}$
Electric charge	coulomb (C)	$A\ s$
Electric potential difference	volt (V)	$kg\ m^2\ s^{-3}\ A^{-1} = J\ A^{-1}\ s^{-1}$
Electric resistance	ohm (Ω)	$kg\ m^2\ s^{-3}\ A^{-2} = V\ A^{-1}$

III. Prefixes[a]

	Prefix	Symbol		Prefix	Symbol
10^{18}	exa	E	10^{-1}	deci	d
10^{15}	peta	P	10^{-2}	centi	c
10^{12}	tera	T	10^{-3}	milli	m
10^{9}	giga	G	10^{-6}	micro	μ
10^{6}	mega	M	10^{-9}	nano	n
10^{3}	kilo	k	10^{-12}	pico	p
10^{2}	hecto	h	10^{-15}	femto	f
10^{1}	deka	D	10^{-18}	atto	a
			10^{-21}	yocto	y

[a] Use of the prefixes hecto, deka, deci, and centi is discouraged.

IV. SI Equivalents of Some Common Non-SI Units

Physical Quantity	Name	SI Equivalent
Length	angstrom (Å)	10^{-10} m
	micron (μ)	10^{-6} m
Volume	liter (L)	10^{-3} m^3
Force	dyne (dyn)	10^{-5} N
Pressure	atmosphere (atm)	101,325 N m^{-2}
	mm Hg (Torr)	133.322 N m^{-2}
Energy	calorie (cal)	4.1840 J
	electron volt (eV)	1.6021×10^{-19} J
	wavenumber (cm^{-1})	1.986×10^{-23} J
	degree kelvin (K)	1.3804×10^{-23} J
	hertz (Hz)	6.6251×10^{-34} J

REFERENCES

Anon. *J. Chem. Educ.* **1971**, *48*, 569.

Heslop, R. B.; Wild, G. M. *SI Units in Chemistry*, Applied Science Publishers Ltd.: London, 1971.

Norris, A. C. *J. Chem. Educ.* **1971**, *48*, 797.

Paul, M. A. *J. Chem. Document.* **1971**, *11*, 3.

Quickenden, T. I.; Marshall, R. C. *J. Chem. Educ.* **1972**, *49*, 114. (Also see Hoppeé, J. I. *J. Chem. Educ.* **1972**, *49*, 505.)

Selected Spectra

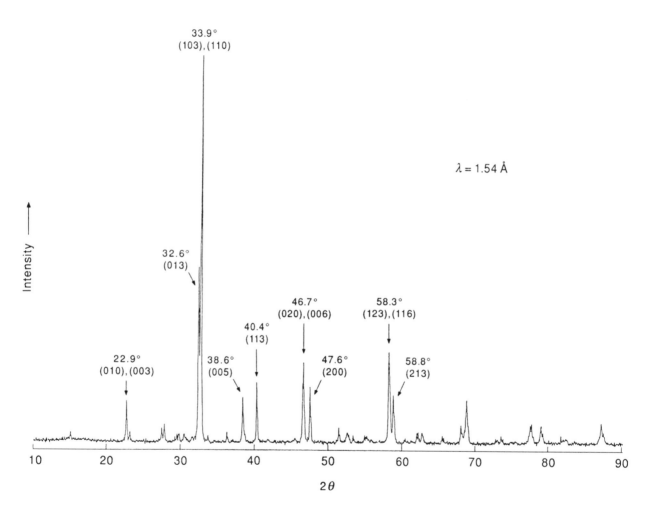

Powder X-ray diffraction pattern for $YBa_2Cu_3O_7$ (Experiment 1)

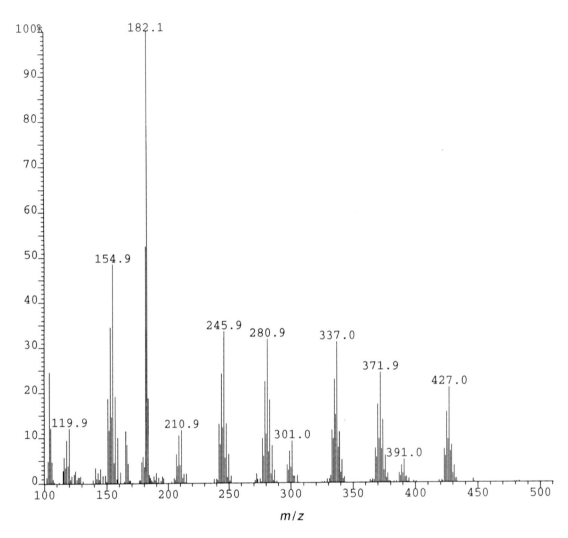

70 eV mass spectrum of $SnCl(CH_2C_6H_5)_3$ (Experiment 7)

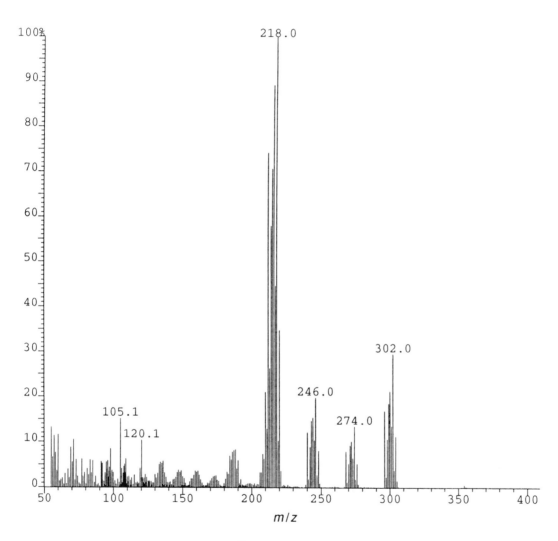

70 eV mass spectrum of $[\eta^6 - C_6H_3(CH_3)_3]Mo(CO)_3$ (Experiment 16)

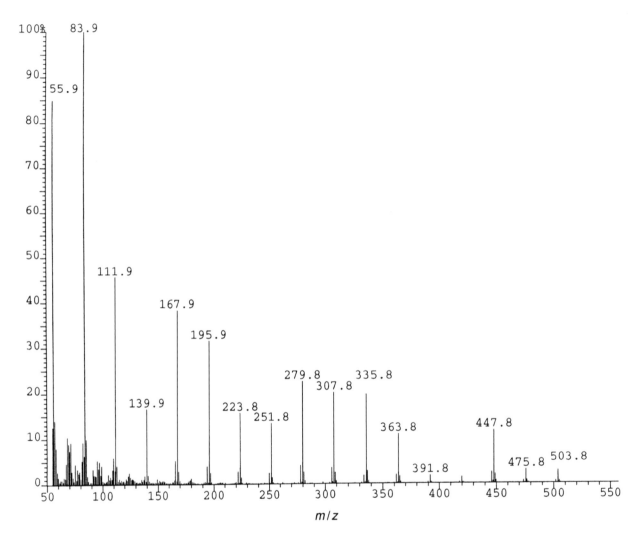

70 eV mass spectrum of $Fe_3(CO)_{12}$ (Experiment 18)

Index

Page numbers followed by (t) indicate tables.